Cram101 Textbook Outlines to accompany:

# Perspectives on Animal Behavior

## Goodenough & McGuire & Wallace, 2nd Edition

An Academic Internet Publishers (AIPI) publication (c) 2007.

Cram101 and Cram101.com are AIPI publications and services. All notes, highlights, reviews, and practice tests are prepared by AIPI for use in AIPI publications, all rights reserved.

You have a discounted membership at www.Cram101.com with this book.

Get all of the practice tests for the chapters of this textbook, and access in-depth reference material for writing essays and papers. Here is an example from a Cram101 Biology text:

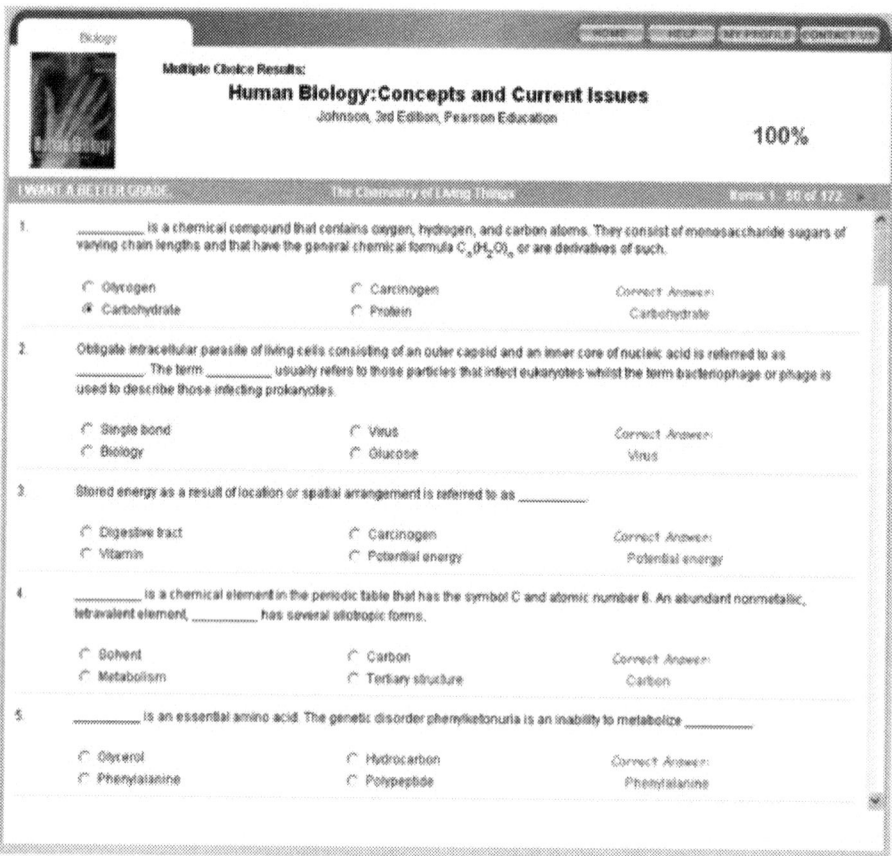

When you need problem solving help with math, stats, and other disciplines, www.Cram101.com will walk through the formulas and solutions step by step.

## With Cram101.com online, you also have access to extensive reference material.

You will nail those essays and papers. Here is an example from a Cram101 Biology text:

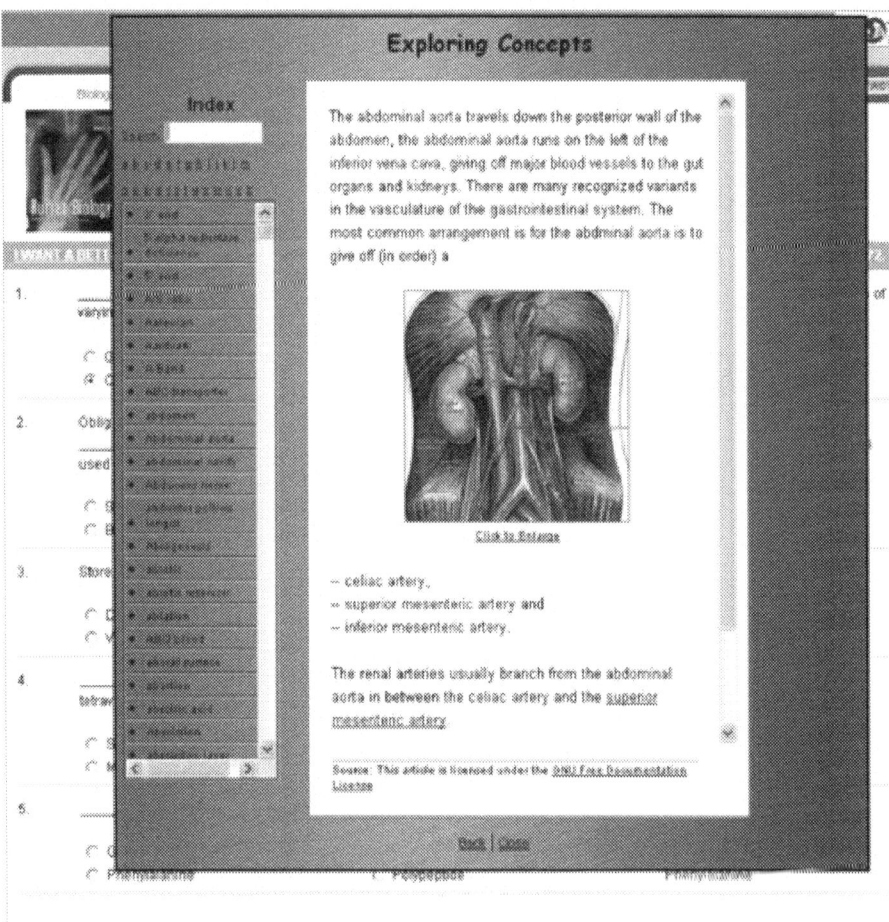

Visit **www.Cram101.com**, click Sign Up at the top of the screen, and enter DK73DW1045 in the promo code box on the registration screen. Access to www.Cram101.com is normally $9.95, but because you have purchased this book, your access fee is only $4.95. Sign up and stop highlighting textbooks forever.

## Learning System

Cram101 Textbook Outlines is a learning system. The notes in this book are the highlights of your textbook, you will never have to highlight a book again.

**How to use this book.** Take this book to class, it is your notebook for the lecture. The notes and highlights on the left hand side of the pages follow the outline and order of the textbook. All you have to do is follow along while your intructor presents the lecture. Circle the items emphasized in class and add other important information on the right side. With Cram101 Textbook Outlines you'll spend less time writing and more time listening. Learning becomes more efficient.

## Cram101.com Online

Increase your studying efficiency by using Cram101.com's practice tests and online reference material. It is the perfect complement to Cram101 Textbook Outlines. Use self-teaching matching tests or simulate in-class testing with comprehensive multiple choice tests, or simply use Cram's true and false tests for quick review. Cram101.com even allows you to enter your in-class notes for an integrated studying format combining the textbook notes with your class notes.

Visit **www.Cram101.com**, click Sign Up at the top of the screen, and enter **DK73DW1045** in the promo code box on the registration screen. Access to www.Cram101.com is normally $9.95, but because you have purchased this book, your access fee is only $4.95. Sign up and stop highlighting textbooks forever.

Copyright © 2008 by Academic Internet Publishers, Inc. All rights reserved. "Cram101"® and "Never Highlight a Book Again!"® are registered trademarks of Academic Internet Publishers, Inc. The Cram101 Textbook Outline series is printed in the United States. ISBN: 1-4288-1937-1

Perspectives on Animal Behavior
Goodenough & McGuire & Wallace, 2nd

# CONTENTS

1. Introduction ................. 2
2. History of the Study of Animal Behavior ................. 6
3. Genetic Analysis of Behavior ................. 16
4. Natural Selection and Ecological Analysis of Behavior ................. 24
5. Learning ................. 34
6. Physiological Analysis Nerve Cells and Behavior ................. 42
7. Physiological Analysis of Behavior - The Endocrine System ................. 52
8. The Development of Behavior ................. 62
9. Biological Clocks ................. 72
10. Mechanisms of Orientation ................. 84
11. The Ecology and Evolution of Spatial Distribution ................. 96
12. Foraging ................. 104
13. Antipredator Behavior ................. 116
14. Sexual Selection ................. 126
15. Parental Care and Mating Systems ................. 134
16. Sociality, Conflict, and Resolution ................. 142
17. Cooperation and Altruism 383 What Is Altruism? ................. 150
18. Maintaining Group Cohesion ................. 160
19. Maintaining Group Cohesion: The Evolution of Communication ................. 170

# Chapter 1. Introduction

| | |
|---|---|
| Hypothesis | In science, an explanation set forth in a manner that can be tested and is capable of being disproved. A tested hypothesis is accepted until and unless it has been disproved. |
| Prey | Organism that is captured and serves as a source of food for an organism of another species is called prey. |
| Well | Well refers to a hole, generally cylindrical and usually walled or lined with pipe, that is dug or drilled into the ground to penetrate an aquifer below the zone of saturation. |
| Species | Group of organisms that resemble one another in appearance, behavior, chemical makeup and processes, and genetic structure is a species. Organisms that reproduce sexually are classified as members of the same species only if they can breed with one another and produce offspring. |
| Extinction | Extinction refers to complete disappearance of a species from the earth. This happens when a species cannot adapt and successfully reproduce under new environmental conditions or when it evolves into one or more new species. Compare speciation. |
| Theory | A general explanation of a characteristic of nature consistently supported by observation or experiment is referred to as a theory. |
| Observations | Information obtained through one or more of the five senses or through instruments that extend the senses are observations. |
| Game species | Type of wild animal that people hunt or fish for, for sport and recreation and sometimes for food is referred to as game species. |
| Food | General term for organic molecules capable of providing energy to heterotrophs when combined with oxygen during biochemical respiration is called food. |
| Migration | A term that refers to the habit of some animals is a migration. |
| Development | Development refers to change from a society that is largely rural, agricultural, illiterate, and poor, with a rapidly growing population, to one that is mostly urban, industrial, educated, and wealthy, with a slowly growing or stationary population. |
| Response | The amount of health damage caused by exposure to a certain dose of a harmful substance or form of radiation is a response. |
| Carnivore | Carnivore refers to animal that feeds on other animals. Compare herbivore, omnivore. |
| Head | A comparatively high promontory with either a CLIFF or steep face. It extends into a large body of water, such as a sea or lake. An unnamed HEAD is usually called a headland. The section of RIP CURRENT which has widened out seaward of the BREAKERS, also called head of rip. |
| Savanna | Savanna refers to an area with trees scattered widely among dense grasses. |
| Hierarchy | Hierarchy refers to grouping of objects by degrees of complexity, grade, or class. A hierarchical system of nomenclature is based on distinctions within groups and between groups. |
| Reproduction | Production of offspring by one is called reproduction. |
| Host | Plant or animal on which a parasite feeds is referred to as host. |
| Degree | An arbitrary measure of temperature. One degree Celsius _ 1.8 degrees Fahrenheit. |
| Benefits | Benefits refers to the economic value of a scheme, usually measured in terms of the cost of damages avoided by the scheme, or the valuation of perceived amenity or environmental improvements. |
| Disperse | Disperse refers to to spread or distribute from a fixed or constant source. To cause to become widely separated. |
| Predation | Predation refers to a situation in which an organism of one species captures and feeds on parts or all of an organism of another species. |

# Chapter 1. Introduction

## Chapter 1. Introduction

| | |
|---|---|
| Genes | Genes refers to coded units of information about specific traits that are passed on from parents to offspring during reproduction. They consist of segments of DNA molecules found in chromosomes. |
| Resources | Resources refer to substances that can be consumed by an organism and, as a result, become unavailable to other organisms. |
| Dna | Large molecules in the cells of organisms that carry genetic information in living organisms are dna. |
| Point | Point refers to the extreme end of a cape, or the outer end of any land area protruding into the water, usually less prominent than a cape. A low profile shoreline promontory of more or less triangular shape, the top of which extends seaward. |
| Risk | Risk refers to the probability that something undesirable will happen from deliberate or accidental exposure. |
| Lead | Lead refers to a heavy metal that is an important constituent of automobile batteries and other industrial products. A toxic metal capable of causing environmental disruption and producing a health problem to people and other living organisms. |
| Event | Event refers to an occurrence meeting specified conditions, e.g. damage, a threshold wave height or a threshold water level. |
| Predator | Predator refers to an organism that captures and feeds on parts or all of an organism of another species. |
| Reach | Reach refers to an arm of the ocean extending into the land. A straight section of restricted waterway of considerable extent; may be similar to a narrows, except much longer in extent. |
| Reduce | With respect to waste management, reduce refers to practices that will reduce the amount of waste we produce. |
| Probability | A mathematical statement about how likely it is that something will happen is probability. |
| State | State refers to an expression of the internal form of matter. Water exists in three states: solid, liquid, and gas. A solid has a fixed volume and fixed shape; a liquid has a fixed volume but no fixed shape; and a gas has neither fixed volume nor fixed shape. |
| Tests | The skeleton or shells of certain microorganisms are called tests. |

# Chapter 1. Introduction

## Chapter 2. History of the Study of Animal Behavior

| | |
|---|---|
| Current | Current refers to the flowing of water, or other liquid or gas. That portion of a stream of water which is moving with a velocity much greater than the average or in which the progress of the water is principally concentrated. Ocean currents can be classified in a number of different ways. |
| Ecology | Ecology refers to study of the interactions of living organisms with one another and with their nonliving environment of matter and energy; study of the structure and functions of nature. |
| Development | Development refers to change from a society that is largely rural, agricultural, illiterate, and poor, with a rapidly growing population, to one that is mostly urban, industrial, educated, and wealthy, with a slowly growing or stationary population. |
| Key | Key refers to a low, insular BANK of sand, coral, etc., as one of the islets off the southern coast of Florida. |
| Theory of evolution | Theory of evolution refers to a widely accepted idea that all life forms developed from earlier life forms. Although this theory conflicts with the creation stories of most religions, it is the way biologists explain how life has changed over the past 3.6-3.8 billion years. |
| Species | Group of organisms that resemble one another in appearance, behavior, chemical makeup and processes, and genetic structure is a species. Organisms that reproduce sexually are classified as members of the same species only if they can breed with one another and produce offspring. |
| System | A set of components that function and interact in some regular and theoretically predictable manner is called a system. |
| Degree | An arbitrary measure of temperature. One degree Celsius _ 1.8 degrees Fahrenheit. |
| Point | Point refers to the extreme end of a cape, or the outer end of any land area protruding into the water, usually less prominent than a cape. A low profile shoreline promontory of more or less triangular shape, the top of which extends seaward. |
| Natural selection | Process by which a particular beneficial gene is reproduced more than other genes in succeeding generations is natural selection. The result of natural selection is a population that contains a greater proportion of organisms better adapted to certain environments. |
| Reproduction | Production of offspring by one is called reproduction. |
| Morphology | Morphology refers to river/estuary/lake/seabed form and its change with time. |
| Population | Group of individual organisms of the same species living within a particular area is referred to as a population. |
| Observations | Information obtained through one or more of the five senses or through instruments that extend the senses are observations. |
| Science | Attempts to discover order in nature and use that knowledge to make predictions about what should happen in nature is called science. |
| Lead | Lead refers to a heavy metal that is an important constituent of automobile batteries and other industrial products. A toxic metal capable of causing environmental disruption and producing a health problem to people and other living organisms. |
| Carnivores | Carnivores refers to organisms that feed on other live organisms; usually applied to animals that eat other animals. |
| Ocean | The great body of salt water which occupies two-thirds of the surface of the Earth, or one of its major subdivisions is called an ocean. |

## Chapter 2. History of the Study of Animal Behavior

## Chapter 2. History of the Study of Animal Behavior

| | |
|---|---|
| Habitat | The place where an organism lives is called habitat. |
| Range | Land used for grazing is referred to as the range. |
| Recent | A synonym of Holocene is called recent. |
| Organism | Any form of life is referred to as an organism. |
| Vertebrates | Vertebrates refer to animals with backbones. Compare with invertebrates. |
| Vertebrate | A chordate with a segmented backbone is a vertebrate. |
| Aquatic | Aquatic refers to pertaining to water. |
| Prey | Organism that is captured and serves as a source of food for an organism of another species is called prey. |
| Food | General term for organic molecules capable of providing energy to heterotrophs when combined with oxygen during biochemical respiration is called food. |
| Situation | The relative geographic location of a site that makes it a good location for a city is a situation. |
| Response | The amount of health damage caused by exposure to a certain dose of a harmful substance or form of radiation is a response. |
| Conditions | Conditions refers to physical or chemical attributes of the environment that, while not being consumed, influence biological processes and population growth. Examples are temperature, salinity, and acidity. Compare resources. |
| Map | Map refers to a representation of Earth's surface usually depicting mostly land areas. |
| Environment | Environment refers to all external conditions and factors, living and nonliving, that affect an organism or other specified system during its lifetime; the earth's life-support systems for us and for all other forms of life-another term for solar capita. |
| Head | A comparatively high promontory with either a CLIFF or steep face. It extends into a large body of water, such as a sea or lake. An unnamed HEAD is usually called a headland. The section of RIP CURRENT which has widened out seaward of the BREAKERS, also called head of |
| Spring | A place where groundwater flows out onto the surface is a spring. |
| State | State refers to an expression of the internal form of matter. Water exists in three states: solid, liquid, and gas. A solid has a fixed volume and fixed shape; a liquid has a fixed volume but no fixed shape; and a gas has neither fixed volume nor fixed shape. |
| Topple | Topple refers to a large rock mass that has fallen over. |
| Shore | That strip of ground bordering any body of water which is alternately exposed, or covered by tides and/or waves is a shore. A shore of unconsolidated material is usually called a beach. |
| Parasite | Consumer organism that lives on or in and feeds on a living plant or animal, known as the host, over an extended period of time is a parasite. The parasite draws nourishment from and gradually weakens its host; it may or may not kill the host. |
| Host | Plant or animal on which a parasite feeds is referred to as host. |
| Event | Event refers to an occurrence meeting specified conditions, e.g. damage, a threshold wave height or a threshold water level. |
| Entrance | The entrance to a navigable BAY, HARBOR or CHANNEL, INLET or mouth separating the ocean from an inland water body. |
| Dorsal | The upper or back surface of an animal is a dorsal. |

Go to Cram101.com for the Practice Tests for this Chapter.

## Chapter 2. History of the Study of Animal Behavior

## Chapter 2. History of the Study of Animal Behavior

| | |
|---|---|
| Base | A substance that combines with a hydrogen ion in solution is called the base. |
| Energy | Capacity to do work by performing mechanical, physical, chemical, or electrical tasks or to cause a heat transfer between two objects at different temperatures is an energy. |
| Experiment | Experiment refers to procedure a scientist uses to study some phenomenon under known conditions. Some experiments are conducted in the laboratory, but others are conducted in nature. The resulting scientific data or facts must be verified or confirmed by repeated observation. |
| Tension | A state of stress that tends to pull the body apart is called tension. |
| Force | Force refers to a push or pull that affects motion. The product of mass and acceleration of a material. |
| Valve | Valve refers to the shell or shells of certain organisms, such as snails and clams. |
| Forcing | With respect to global change, processes capable of changing global temperature, such as changes in solar energy emitted from the sun, or volcanic activity, we have forcing. |
| Water level | Elevation of a particular point or small patch on the surface of a body of water above a specific point or surface, averaged over a period of time sufficiently long to remove the effects of short period disturbances is referred to as water level. |
| Trough | A long and broad submarine depression with gently sloping sides is referred to as a trough. |
| Threshold | Threshold refers to a point in the operation of a system at which a change occurs. With respect to toxicology, it is a level below which effects are not observable and above which effects become apparent. |
| Reach | Reach refers to an arm of the ocean extending into the land. A straight section of restricted waterway of considerable extent; may be similar to a narrows, except much longer in extent. |
| Caudal fin | The tail of a fish is referred to as caudal fin. |
| Probability | A mathematical statement about how likely it is that something will happen is probability. |
| Well | Well refers to a hole, generally cylindrical and usually walled or lined with pipe, that is dug or drilled into the ground to penetrate an aquifer below the zone of saturation. |
| Loop | That part of a STANDING WAVE where the vertical motion is greatest and the horizontal velocities are least is a loop. |
| Law | A large construct explaining events in nature that have been observed to occur with unvarying uniformity under the same conditions is called law. |
| Planet | A smaller, usually nonluminous body orbiting a star is a planet. |
| Frequency | Number of events in a given time interval. For earthquakes, it is the number of cycles of seismic waves that pass in a second; frequency = l/period. |
| Matter | Matter refers to anything that has mass and takes up space. On earth, where gravity is present, we weigh an object to determine its mass. |
| Extinction | Extinction refers to complete disappearance of a species from the earth. This happens when a species cannot adapt and successfully reproduce under new environmental conditions or when it evolves into one or more new species. Compare speciation. |
| Hemispheres | Hemispheres refers to earth parallel to the equator. The equator is 0° latitude and other lines arc proportioned up to 90°N or 90°S. Perpendicular to longitude. |
| Tests | The skeleton or shells of certain microorganisms are called tests. |
| Coast | A strip of land of indefinite length and width that extends from the SEASHORE inland to the |

# Chapter 2. History of the Study of Animal Behavior

## Chapter 2. History of the Study of Animal Behavior

| | |
|---|---|
| | first major change in terrain features is referred to as coast. |
| Constancy | Ability of a living system, such as a population, to maintain a certain size. Compare inertia, resilienc is called constancy. |
| Genes | Genes refers to coded units of information about specific traits that are passed on from parents to offspring during reproduction. They consist of segments of DNA molecules found in chromosomes. |
| Front | The boundary between two air masses with different temperatures and densitie is referred to as front. |
| Interference | Interference refers to during succession, one species prevents the entrance of later successional species into an ecosystem. For example, some grasses produce such dense and thick mats that seeds of trees cannot reach the soil to germinate. As long as these grasses persist, th. |
| Bank | The rising ground bordering a lake, river or sea is referred to as a bank. |
| Dip | The angle of inclination measured in degrees from the horizontal is called dip. |
| Stream | Stream refers to any flow of water; a current. A course of water flowing along a bed in the earth. |
| Ecological niche | Total way of life or role of a species in an ecosystem. It includes all physical, chemical, and biological conditions a species needs to live and reproduce in an ecosyste is referred to as the ecological niche. |
| Tongue | A long narrow strip of land, projecting into a body of water is a tongue. |
| Strain | A change in torn or size of a body due to external forces is called strain. |
| Cliff | A high steep face of rock is referred to as cliff. |
| Divergence | Divergence refers to to move apart from a common source. |
| Resources | Resources refer to substances that can be consumed by an organism and, as a result, become unavailable to other organisms. |
| Benefits | Benefits refers to the economic value of a scheme, usually measured in terms of the cost of damages avoided by the scheme, or the valuation of perceived amenity or environmental improvements. |
| Predator | Predator refers to an organism that captures and feeds on parts or all of an organism of another species. |
| Landmark | Landmark refers to a conspicious object, natural or man-made, located near or on land, which aids in fixing the position of an observer. |
| Community | Community refers to populations of all species living and interacting in an area at a particular time. |
| Age structure | Percentage of the population at each age level in a population is referred to as the age structure. |
| Theories | Scientific models that offer broad, fundamental explanations of related phenomena and are supported by consistent and extensive evidence are theories. |
| Ebb | Period when tide level is falling; often taken to mean the EBB CURRENT which occurs during this period. |
| Coral | Coral refers to any of more than 6,000 species of small cnidarians, many of which are capable of generating hard calcareous skeletons. |

## Chapter 2. History of the Study of Animal Behavior

| | |
|---|---|
| **Hypothesis** | In science, an explanation set forth in a manner that can be tested and is capable of being disproved. A tested hypothesis is accepted until and unless it has been disproved. |
| **Site** | A factor considering the summation of all environmental features of a location that influences the placement of a city is a site. |
| **Invertebrates** | Invertebrates refers to animals that have no backbones. Compare vertebrates. |
| **Cell** | Smallest living unit of an organism. Each cell is encased in an outer membrane or wall and contains genetic material and other parts to perform its life function. Organisms such as bacteria consist of only one cell, but most of the organisms we are. |

## Chapter 3. Genetic Analysis of Behavior

| | |
|---|---|
| Genes | Genes refers to coded units of information about specific traits that are passed on from parents to offspring during reproduction. They consist of segments of DNA molecules found in chromosomes. |
| Gene | Gene refers to the fundamental unit in inheritance; it carries the characteristics of parents to their offspring. |
| Mutation | A random change in DNA molecules making up genes that can yield changes in anatomy, physiology, or behavior in offspring is a mutation. |
| System | A set of components that function and interact in some regular and theoretically predictable manner is called a system. |
| Probability | A mathematical statement about how likely it is that something will happen is probability. |
| Frequency | Number of events in a given time interval. For earthquakes, it is the number of cycles of seismic waves that pass in a second; frequency = l/period. |
| Population | Group of individual organisms of the same species living within a particular area is referred to as a population. |
| Protein | Protein refers to a family of complex organic compounds containing nitrogen and composed of various amino acids. |
| Organism | Any form of life is referred to as an organism. |
| Chemical | One of the millions of different elements and compounds found naturally or synthesized by human is referred to as chemical. |
| Response | The amount of health damage caused by exposure to a certain dose of a harmful substance or form of radiation is a response. |
| Dna | Large molecules in the cells of organisms that carry genetic information in living organisms are dna. |
| Acid | A substance that releases a hydrogen ion in solution is an acid. |
| Phosphate | An essential plant macronutrient is phosphate. |
| Base | A substance that combines with a hydrogen ion in solution is called the base. |
| Molecule | Molecule refers to a combination of two or more atoms of the same chemical element or different chemical elements held together by chemical bonds. Compare with atom, ion. |
| Messenger | A free-falling weight that is attached to a cable and used to trigger the closure of water-sampling bottles is referred to as a messenger. |
| Environment | Environment refers to all external conditions and factors, living and nonliving, that affect an organism or other specified system during its lifetime; the earth's life-support systems for us and for all other forms of life-another term for solar capita. |
| Cell | Smallest living unit of an organism. Each cell is encased in an outer membrane or wall and contains genetic material and other parts to perform its life function. Organisms such as bacteria consist of only one cell, but most of the organisms we are. |
| Vertebrates | Vertebrates refer to animals with backbones. Compare with invertebrates. |
| Concentration | Amount of a chemical in a particular volume or weight of air, water, soil, or other medium is referred to as concentration. |
| Input | Input refers to matter, energy, or information entering a system. Compare output, throughput. |
| Species | Group of organisms that resemble one another in appearance, behavior, chemical makeup and processes, and genetic structure is a species. Organisms that reproduce sexually are |

# Chapter 3. Genetic Analysis of Behavior

## Chapter 3. Genetic Analysis of Behavior

| | |
|---|---|
| | classified as members of the same species only if they can breed with one another and produce offspring. |
| Experiment | Experiment refers to procedure a scientist uses to study some phenomenon under known conditions. Some experiments are conducted in the laboratory, but others are conducted in nature. The resulting scientific data or facts must be verified or confirmed by repeated observation. |
| Energy | Capacity to do work by performing mechanical, physical, chemical, or electrical tasks or to cause a heat transfer between two objects at different temperatures is an energy. |
| Hypothesis | In science, an explanation set forth in a manner that can be tested and is capable of being disproved. A tested hypothesis is accepted until and unless it has been disproved. |
| Density | Density refers to the ratio of a mass to a unit volume specified as grams per cubic centimeter. |
| Chromosome | A grouping of various genes and associated proteins in plant and animal cells that carry certain types of genetic informatio are called the chromosome. |
| Conditions | Conditions refers to physical or chemical attributes of the environment that, while not being consumed, influence biological processes and population growth. Examples are temperature, salinity, and acidity. Compare resources. |
| Strain | A change in torn or size of a body due to external forces is called strain. |
| Genetic diversity | Genetic diversity refers to variability in the genetic makeup among individuals within a single species. |
| Alleles | Slightly different molecular forms found in a particular gene are alleles. |
| Matter | Matter refers to anything that has mass and takes up space. On earth, where gravity is present, we weigh an object to determine its mass. |
| Scientific method | An investigative technique whereby data are collected objectively, interpreted as a hypothesis, and then tested by falsification is referred to as scientific method. |
| Predator | Predator refers to an organism that captures and feeds on parts or all of an organism of another species. |
| Predation | Predation refers to a situation in which an organism of one species captures and feeds on parts or all of an organism of another species. |
| Larva | The immature form of an animal that differs significantly from the adult form is a larva. |
| Caudal fin | The tail of a fish is referred to as caudal fin. |
| Food | General term for organic molecules capable of providing energy to heterotrophs when combined with oxygen during biochemical respiration is called food. |
| Natural selection | Process by which a particular beneficial gene is reproduced more than other genes in succeeding generations is natural selection. The result of natural selection is a population that contains a greater proportion of organisms better adapted to certain environments. |
| Temperature | Temperature refers to a measure of the average speed of motion of the atoms, ions, or molecules in a substance or combination of substances at a given moment. Compare with heat. |
| Lead | Lead refers to a heavy metal that is an important constituent of automobile batteries and other industrial products. A toxic metal capable of causing environmental disruption and producing a health problem to people and other living organisms. |
| Frequency distribution | The arrangement of data that shows the occurrence and range of the values of a variable is a frequency distribution. |

# Chapter 3. Genetic Analysis of Behavior

## Chapter 3. Genetic Analysis of Behavior

| | |
|---|---|
| Population density | Number of organisms in a particular population found in a specified area is the population density. Compare with ecological population density. |
| Output | Output refers to matter, energy, or information leaving a system. Compare with input, throughput. |
| Event | Event refers to an occurrence meeting specified conditions, e.g. damage, a threshold wave height or a threshold water level. |
| Well | Well refers to a hole, generally cylindrical and usually walled or lined with pipe, that is dug or drilled into the ground to penetrate an aquifer below the zone of saturation. |
| Science | Attempts to discover order in nature and use that knowledge to make predictions about what should happen in nature is called science. |
| Key | Key refers to a low, insular BANK of sand, coral, etc., as one of the islets off the southern coast of Florida. |
| Sea | Sea refers to the ocean. A large body of salt water, second in rank to an ocean, more or less landlocked and generally part of, or connected with, an ocean or a larger sea. State of the ocean or lake surface, in regard to waves. |
| Site | A factor considering the summation of all environmental features of a location that influences the placement of a city is a site. |
| Gut | Gut refers to a narrow passage such as a strait or INLET. A CHANNEL in otherwise shallow water, generally formed by water in motion. |
| Power | Power refers to the time rate of doing work. |
| Point | Point refers to the extreme end of a cape, or the outer end of any land area protruding into the water, usually less prominent than a cape. A low profile shoreline promontory of more or less triangular shape, the top of which extends seaward. |
| Nucleus | Extremely tiny center of an atom, making up most of the atom's mass is the nucleus. It contains one or more positively charged protons and one or more neutrons with no electrical charge. |
| Development | Development refers to change from a society that is largely rural, agricultural, illiterate, and poor, with a rapidly growing population, to one that is mostly urban, industrial, educated, and wealthy, with a slowly growing or stationary population. |
| Migration | A term that refers to the habit of some animals is a migration. |
| Cross section | A two-dimensional drawing showing features in the vertical plane as in a canvon wall or road cut is referred to as cross section. |
| Map | Map refers to a representation of Earth's surface usually depicting mostly land areas. |
| Head | A comparatively high promontory with either a CLIFF or steep face. It extends into a large body of water, such as a sea or lake. An unnamed HEAD is usually called a headland. The section of RIP CURRENT which has widened out seaward of the BREAKERS, also called head of |
| Ventral | Ventral refers to of or pertaining to the underside of animals. |
| Dorsal | The upper or back surface of an animal is a dorsal. |
| Mixture | Mixture refers to combination of two or more elements and compounds. |
| Recent | A synonym of Holocene is called recent. |
| Pool | Common bed form produced by scour in meandering and straight channels is a pool. |
| Tests | The skeleton or shells of certain microorganisms are called tests. |

## Chapter 3. Genetic Analysis of Behavior

| | |
|---|---|
| **Fold** | Fold refers to wavy geologic structures formed by the compression and bending of sedimentary layers. |
| **Mass** | The amount of material in an object is the mass. |
| **Rock** | Any material that makes up a large, natural, continuous part of earth's crust is rock. |
| **Wave** | An oscillatory movement in a body of water manifested by an alternate rise and fall of the surface is called a wave. Disturbances of the surface of a liquid body, as the ocean, in the form of a ridge, swell or hump. The term wave by itself usually refers to the term surface gravity wave. |
| **Strand** | Strand refers to shore or beach of the ocean or a large lake. The land bordering any large body of water, especially a sea or an arm of the ocean. |
| **Electron** | Tiny particle moving around outside the nucleus of an atom. Each electron has one unit of negative charge and almost no mass. Compare neutron, proton. |
| **Ion** | Ion refers to atom or group of atoms with one or more positive or negative electrical charges. Compare atone, molecule. |

## Chapter 4. Natural Selection and Ecological Analysis of Behavior

| | |
|---|---|
| Natural selection | Process by which a particular beneficial gene is reproduced more than other genes in succeeding generations is natural selection. The result of natural selection is a population that contains a greater proportion of organisms better adapted to certain environments. |
| Adaptation | Any genetically controlled structural, physiological, or behavioral characteristic that helps an organism survive and reproduce under a given set of environmental conditions. It usually results from a beneficial mutation is an adaptation. |
| Key | Key refers to a low, insular BANK of sand, coral, etc., as one of the islets off the southern coast of Florida. |
| Gene pool | Gene pool refers to the sum total of all genes found in the individuals of the population of a particular species. |
| Pool | Common bed form produced by scour in meandering and straight channels is a pool. |
| Alleles | Slightly different molecular forms found in a particular gene are alleles. |
| Genes | Genes refers to coded units of information about specific traits that are passed on from parents to offspring during reproduction. They consist of segments of DNA molecules found in chromosomes. |
| Population | Group of individual organisms of the same species living within a particular area is referred to as a population. |
| Frequency | Number of events in a given time interval. For earthquakes, it is the number of cycles of seismic waves that pass in a second; frequency = l/period. |
| Matter | Matter refers to anything that has mass and takes up space. On earth, where gravity is present, we weigh an object to determine its mass. |
| Food | General term for organic molecules capable of providing energy to heterotrophs when combined with oxygen during biochemical respiration is called food. |
| Habitat | The place where an organism lives is called habitat. |
| Species | Group of organisms that resemble one another in appearance, behavior, chemical makeup and processes, and genetic structure is a species. Organisms that reproduce sexually are classified as members of the same species only if they can breed with one another and produce offspring. |
| Competition | Two or more individual organisms of a single species or two or more individuals of different species attempting to use the same scarce resources in the same ecosystem is a competition. |
| Point | Point refers to the extreme end of a cape, or the outer end of any land area protruding into the water, usually less prominent than a cape. A low profile shoreline promontory of more or less triangular shape, the top of which extends seaward. |
| Preservation | Static protection of an area or element, attempting to perpetuate the existence of a given 'state' is called preservation. |
| Conditions | Conditions refers to physical or chemical attributes of the environment that, while not being consumed, influence biological processes and population growth. Examples are temperature, salinity, and acidity. Compare resources. |
| Recent | A synonym of Holocene is called recent. |
| Observations | Information obtained through one or more of the five senses or through instruments that extend the senses are observations. |
| Resources | Resources refer to substances that can be consumed by an organism and, as a result, become unavailable to other organisms. |

# Chapter 4. Natural Selection and Ecological Analysis of Behavior

## Chapter 4. Natural Selection and Ecological Analysis of Behavior

| | |
|---|---|
| **Differential reproduction** | Phenomenon in which individuals with adaptive genetic traits produce more living offspring than do individuals without such trait is a differential reproduction. |
| **Reproduction** | Production of offspring by one is called reproduction. |
| **Lead** | Lead refers to a heavy metal that is an important constituent of automobile batteries and other industrial products. A toxic metal capable of causing environmental disruption and producing a health problem to people and other living organisms. |
| **Environment** | Environment refers to all external conditions and factors, living and nonliving, that affect an organism or other specified system during its lifetime; the earth's life-support systems for us and for all other forms of life-another term for solar capita. |
| **Gradient** | A measure of slope in meters of rise or fall per meter of horizontal distance. More general, a change of a value per unit of distance, e.g. the GRADIENT in longshore transport causes EROSION or ACCRETION. With reference to winds or currents, the rate of increase or decrease in speed, usually in the vertical; or the curve that represents this rate. |
| **Fall** | Fall refers to a mass moving nearly vertical and downward under the influence of gravity. |
| **Mutation** | A random change in DNA molecules making up genes that can yield changes in anatomy, physiology, or behavior in offspring is a mutation. |
| **Chromosome** | A grouping of various genes and associated proteins in plant and animal cells that carry certain types of genetic informatio are called the chromosome. |
| **Cell** | Smallest living unit of an organism. Each cell is encased in an outer membrane or wall and contains genetic material and other parts to perform its life function. Organisms such as bacteria consist of only one cell, but most of the organisms we are. |
| **Protein** | Protein refers to a family of complex organic compounds containing nitrogen and composed of various amino acids. |
| **Genetic diversity** | Genetic diversity refers to variability in the genetic makeup among individuals within a single species. |
| **Mixture** | Mixture refers to combination of two or more elements and compounds. |
| **Predation** | Predation refers to a situation in which an organism of one species captures and feeds on parts or all of an organism of another species. |
| **Prey** | Organism that is captured and serves as a source of food for an organism of another species is called prey. |
| **Predator** | Predator refers to an organism that captures and feeds on parts or all of an organism of another species. |
| **Sexual reproduction** | Sexual reproduction refers to reproduction in organisms that produce offspring by combining sex cells or gametes from both parents. This produces offspring that have combinations of traits from their parents. Compare with asexual reproduction. |
| **Density** | Density refers to the ratio of a mass to a unit volume specified as grams per cubic centimeter. |
| **Hypothesis** | In science, an explanation set forth in a manner that can be tested and is capable of being disproved. A tested hypothesis is accepted until and unless it has been disproved. |
| **Tests** | The skeleton or shells of certain microorganisms are called tests. |
| **Glass** | Glass refers to matter created when magma cools too quickly for atoms to arrange themselves into the ordered atomic structures of minerals. Most glasses are supercooled liquids. |
| **Situation** | The relative geographic location of a site that makes it a good location for a city is a |

Go to **Cram101.com** for the Practice Tests for this Chapter.

## Chapter 4. Natural Selection and Ecological Analysis of Behavior

| | |
|---|---|
| | situation. |
| Range | Land used for grazing is referred to as the range. |
| Entrance | The entrance to a navigable BAY, HARBOR or CHANNEL, INLET or mouth separating the ocean from an inland water body. |
| Riparian | Riparian refers to pertaining to the banks of a body of water. |
| Desert | Desert refers to biome in which evaporation exceeds precipitation and the average amount of precipitation is less than 25 centimeters a year. Such areas have little vegetation or have widely spaced, mostly low vegetation. Compare forest, grassland. |
| Grassland | Grassland refers to biome found in regions where moderate annual average precipitation is enough to support the growth of grass and small plants, but not enough to support large stands of trees. Compare desert, forest. |
| Degree | An arbitrary measure of temperature. One degree Celsius _ 1.8 degrees Fahrenheit. |
| River | A natural stream of water larger than a brook or creek is a river. |
| Well | Well refers to a hole, generally cylindrical and usually walled or lined with pipe, that is dug or drilled into the ground to penetrate an aquifer below the zone of saturation. |
| Energy | Capacity to do work by performing mechanical, physical, chemical, or electrical tasks or to cause a heat transfer between two objects at different temperatures is an energy. |
| Law | A large construct explaining events in nature that have been observed to occur with unvarying uniformity under the same conditions is called law. |
| Site | A factor considering the summation of all environmental features of a location that influences the placement of a city is a site. |
| Response | The amount of health damage caused by exposure to a certain dose of a harmful substance or form of radiation is a response. |
| Lava | Molten rock that is extruded out of volcanoes is called lava. |
| Bed | The bottom of a watercourse, or any body of water is called a bed. |
| Thermal | The energy of the random motion of atoms and molecules is referred to as thermal. |
| Gene flow | Movement of genes between populations, which can lead to changes in the genetic composition of local populations is referred to as gene flow. |
| Science | Attempts to discover order in nature and use that knowledge to make predictions about what should happen in nature is called science. |
| Risk | Risk refers to the probability that something undesirable will happen from deliberate or accidental exposure. |
| Offshore | Offshore in beach terminology refers to the comparatively flat zone of variable width, extending from the shoreface to the edge of the continental shelf. It is continually submerged. The direction seaward from the shore. The zone beyond the nearshore zone where sediment motion induced by waves alone effectively ceases and where the influence of the sea bed on wave action is small in comparison with the effect of wind. The breaker zone directly seaward of the low tide line. |
| Reduce | With respect to waste management, reduce refers to practices that will reduce the amount of waste we produce. |
| Shore | That strip of ground bordering any body of water which is alternately exposed, or covered by tides and/or waves is a shore. A shore of unconsolidated material is usually called a beach. |

# Chapter 4. Natural Selection and Ecological Analysis of Behavior

## Chapter 4. Natural Selection and Ecological Analysis of Behavior

| | |
|---|---|
| Cliff | A high steep face of rock is referred to as cliff. |
| Ledge | Ledge refers to a rocky formation continuous with and fringing the shore. |
| Wind | Wind refers to the mass movement of air. |
| Selective pressure | A factor in a population's environment that causes natural selection to occur is selective pressure. |
| Bank | The rising ground bordering a lake, river or sea is referred to as a bank. |
| Forest | Forest refers to biome with enough average annual precipitation to support growth of various species of trees and smaller forms of vegetation. Compare desert, grassland. |
| Bacteria | Bacteria refer to prokaryotic, one-celled organisms. Some transmit diseases. Most act as decomposers and get the nutrients they need by breaking down complex organic compounds in the tissues of living or dead organisms into simpler inorganic nutrient compounds. |
| Benefits | Benefits refers to the economic value of a scheme, usually measured in terms of the cost of damages avoided by the scheme, or the valuation of perceived amenity or environmental improvements. |
| Theory | A general explanation of a characteristic of nature consistently supported by observation or experiment is referred to as a theory. |
| Magnitude | An assessment of the size of an event is a magnitude. Magnitude scales exist for earthquakes, volcanic eruptions. hurricanes, and tornadoes. For earthquakes, different magnitudes are calculated for the same earthquake when different types of seismic waves are used. |
| Spring | A place where groundwater flows out onto the surface is a spring. |
| Solar system | The sun together with the planets and other bodies that revolve around it is a solar system. |
| System | A set of components that function and interact in some regular and theoretically predictable manner is called a system. |
| Gravity | The attraction between bodies of matter is a gravity. |
| Planet | A smaller, usually nonluminous body orbiting a star is a planet. |
| Development | Development refers to change from a society that is largely rural, agricultural, illiterate, and poor, with a rapidly growing population, to one that is mostly urban, industrial, educated, and wealthy, with a slowly growing or stationary population. |
| Rock | Any material that makes up a large, natural, continuous part of earth's crust is rock. |
| Stable equilibrium | Stable equilibrium refers to a condition in which a system will remain if undisturbed and to which it will return when displaced. |
| Equilibrium | A point of rest. A system that does not tend to undergo any change of its own accord but remains in a single, fixed condition is said to be in equilibrium. Compare with steady state. |
| Reach | Reach refers to an arm of the ocean extending into the land. A straight section of restricted waterway of considerable extent; may be similar to a narrows, except much longer in extent. |
| Stability | Ability of a living system to withstand or recover from externally imposed changes or stresses is called stability. |
| Force | Force refers to a push or pull that affects motion. The product of mass and acceleration of a material. |
| Catastrophe | Catastrophe refers to a situation or event that causes significant damage to people and property, such that recovery and/or rehabilitation is a long and involved process. Examples of natural catastrophes include hurricanes, volcanic eruptions, large wildfires, and floods. |

## Chapter 4. Natural Selection and Ecological Analysis of Behavior

| | |
|---|---|
| **Coast** | A strip of land of indefinite length and width that extends from the SEASHORE inland to the first major change in terrain features is referred to as coast. |

# Chapter 4. Natural Selection and Ecological Analysis of Behavior

## Chapter 5. Learning

| | |
|---|---|
| Adaptation | Any genetically controlled structural, physiological, or behavioral characteristic that helps an organism survive and reproduce under a given set of environmental conditions. It usually results from a beneficial mutation is an adaptation. |
| Species | Group of organisms that resemble one another in appearance, behavior, chemical makeup and processes, and genetic structure is a species. Organisms that reproduce sexually are classified as members of the same species only if they can breed with one another and produce offspring. |
| Natural selection | Process by which a particular beneficial gene is reproduced more than other genes in succeeding generations is natural selection. The result of natural selection is a population that contains a greater proportion of organisms better adapted to certain environments. |
| Conditions | Conditions refers to physical or chemical attributes of the environment that, while not being consumed, influence biological processes and population growth. Examples are temperature, salinity, and acidity. Compare resources. |
| Benefits | Benefits refers to the economic value of a scheme, usually measured in terms of the cost of damages avoided by the scheme, or the valuation of perceived amenity or environmental improvements. |
| Environment | Environment refers to all external conditions and factors, living and nonliving, that affect an organism or other specified system during its lifetime; the earth's life-support systems for us and for all other forms of life-another term for solar capita. |
| Degree | An arbitrary measure of temperature. One degree Celsius _ 1.8 degrees Fahrenheit. |
| Ecology | Ecology refers to study of the interactions of living organisms with one another and with their nonliving environment of matter and energy; study of the structure and functions of nature. |
| Food | General term for organic molecules capable of providing energy to heterotrophs when combined with oxygen during biochemical respiration is called food. |
| Spring | A place where groundwater flows out onto the surface is a spring. |
| Sand | Sand refers to an unconsolidated mixture of inorganic soil consisting of small but easily distinguishable grains ranging in size from about .062 mm to 2.0 mm. |
| Accuracy | Accuracy refers to the extent to which a measurement agrees with the accepted or correct value for that quantity, based on careful measurements by many people over a long time. Compare to precision. |
| Experiment | Experiment refers to procedure a scientist uses to study some phenomenon under known conditions. Some experiments are conducted in the laboratory, but others are conducted in nature. The resulting scientific data or facts must be verified or confirmed by repeated observation. |
| Key | Key refers to a low, insular BANK of sand, coral, etc., as one of the islets off the southern coast of Florida. |
| Hypothesis | In science, an explanation set forth in a manner that can be tested and is capable of being disproved. A tested hypothesis is accepted until and unless it has been disproved. |
| Fall | Fall refers to a mass moving nearly vertical and downward under the influence of gravity. |
| Interference | Interference refers to during succession, one species prevents the entrance of later successional species into an ecosystem. For example, some grasses produce such dense and thick mats that seeds of trees cannot reach the soil to germinate. As long as these grasses persist, th. |

## Chapter 5. Learning

| | |
|---|---|
| Site | A factor considering the summation of all environmental features of a location that influences the placement of a city is a site. |
| Recent | A synonym of Holocene is called recent. |
| Tests | The skeleton or shells of certain microorganisms are called tests. |
| Observations | Information obtained through one or more of the five senses or through instruments that extend the senses are observations. |
| Core | Core refers to a cylindrical sample extracted from a beach or seabed to investigate the types and DEPTHS of sediment layers. An inner, often much less permeable portion of a BREAKWATER, or BARRIER beach. |
| Debris | Debris refers to any accumulation of rock fragments; detritus. |
| System | A set of components that function and interact in some regular and theoretically predictable manner is called a system. |
| Situation | The relative geographic location of a site that makes it a good location for a city is a situation. |
| Probability | A mathematical statement about how likely it is that something will happen is probability. |
| Development | Development refers to change from a society that is largely rural, agricultural, illiterate, and poor, with a rapidly growing population, to one that is mostly urban, industrial, educated, and wealthy, with a slowly growing or stationary population. |
| Response | The amount of health damage caused by exposure to a certain dose of a harmful substance or form of radiation is a response. |
| Polychaete | A class of annelid worms that includes most marine segmented worms is called polychaete. |
| Mud | Mud refers to a mixture of silt and clay sized particles. |
| Predator | Predator refers to an organism that captures and feeds on parts or all of an organism of another species. |
| Prey | Organism that is captured and serves as a source of food for an organism of another species is called prey. |
| Glass | Glass refers to matter created when magma cools too quickly for atoms to arrange themselves into the ordered atomic structures of minerals. Most glasses are supercooled liquids. |
| Well | Well refers to a hole, generally cylindrical and usually walled or lined with pipe, that is dug or drilled into the ground to penetrate an aquifer below the zone of saturation. |
| Sea | Sea refers to the ocean. A large body of salt water, second in rank to an ocean, more or less landlocked and generally part of, or connected with, an ocean or a larger sea. State of the ocean or lake surface, in regard to waves. |
| Algae | Algae refers to simple marine and freshwater plants, unicellular and multicellular, that lack roots, stems, and leaves. |
| Energy | Capacity to do work by performing mechanical, physical, chemical, or electrical tasks or to cause a heat transfer between two objects at different temperatures is an energy. |
| Duration | In forecasting waves, the length of time the wind blows in essentially the same direction over the FETCH is a duration. |
| Coast | A strip of land of indefinite length and width that extends from the SEASHORE inland to the first major change in terrain features is referred to as coast. |
| Intertidal zone | The area of shoreline between low and high tides are called the intertidal zone. |

Go to Cram101.com for the Practice Tests for this Chapter.

## Chapter 5. Learning

| | |
|---|---|
| Zone | Division or province of the ocean with homogeneous characteristics is referred to as a zone. |
| Habitat | The place where an organism lives is called habitat. |
| Rocks | An aggregate of one or more minerals rather large in area are rocks. The three classes of rocks are the following: Igneous rock - crystalline rocks formed from molten material. Sedimentary rock - A rock resulting from the consolidation of loose sediment that has accumulated in layers. Metamorphic rock - Rock that has formed from preexisting rock as a result of heat or pressure. |
| Frequency | Number of events in a given time interval. For earthquakes, it is the number of cycles of seismic waves that pass in a second; frequency = l/period. |
| Brine | Brine refers to water having a much higher. |
| Point | Point refers to the extreme end of a cape, or the outer end of any land area protruding into the water, usually less prominent than a cape. A low profile shoreline promontory of more or less triangular shape, the top of which extends seaward. |
| Event | Event refers to an occurrence meeting specified conditions, e.g. damage, a threshold wave height or a threshold water level. |
| Extinction | Extinction refers to complete disappearance of a species from the earth. This happens when a species cannot adapt and successfully reproduce under new environmental conditions or when it evolves into one or more new species. Compare speciation. |
| Resource | Resource refers to anything obtained from the living and nonliving environment to meet human needs and wants. It can also be applied to other species. |
| Chemical | One of the millions of different elements and compounds found naturally or synthesized by human is referred to as chemical. |
| Head | A comparatively high promontory with either a CLIFF or steep face. It extends into a large body of water, such as a sea or lake. An unnamed HEAD is usually called a headland. The section of RIP CURRENT which has widened out seaward of the BREAKERS, also called head of |
| Reach | Reach refers to an arm of the ocean extending into the land. A straight section of restricted waterway of considerable extent; may be similar to a narrows, except much longer in extent. |
| Bar | An offshore ridge or mound of sand, gravel, or other unconsolidated material which is submerged, especially at the mouth of a river or estuary, or lying parallel to, and a short distance from, the beach is referred to as a bar. |
| Clay | Clay refers to a fine grained sediment with a typical grain size less than 0.004 mm. Possesses electromagnetic properties which bind the grains together to give a bulk strength or cohesion. |
| Mass | The amount of material in an object is the mass. |
| Community | Community refers to populations of all species living and interacting in an area at a particular time. |
| Map | Map refers to a representation of Earth's surface usually depicting mostly land areas. |
| Loop | That part of a STANDING WAVE where the vertical motion is greatest and the horizontal velocities are least is a loop. |
| Lead | Lead refers to a heavy metal that is an important constituent of automobile batteries and other industrial products. A toxic metal capable of causing environmental disruption and producing a health problem to people and other living organisms. |
| Reduce | With respect to waste management, reduce refers to practices that will reduce the amount of |

# Chapter 5. Learning

## Chapter 5. Learning

| | |
|---|---|
| | waste we produce. |
| Organism | Any form of life is referred to as an organism. |
| River | A natural stream of water larger than a brook or creek is a river. |
| Lake | Large natural body of standing fresh water formed when water from precipitation, land runoff, or groundwater flow fills a depression in the earth created by glaciation, earth movement, volcanic activity, or a giant meteorit are called the lake. |
| Population | Group of individual organisms of the same species living within a particular area is referred to as a population. |
| Continent | Continent refers to lower-density masses of rock, exposed as about 40 percent of the Earth's surface: 29 percent as land and I 1 percent as the floor of shallow seas. |

# Chapter 5. Learning

## Chapter 6. Physiological Analysis Nerve Cells and Behavior

| | |
|---|---|
| Permeability | Permeability refers to the property of bulk material which permits movement of water through its pores. |
| Prey | Organism that is captured and serves as a source of food for an organism of another species is called prey. |
| Food | General term for organic molecules capable of providing energy to heterotrophs when combined with oxygen during biochemical respiration is called food. |
| Environment | Environment refers to all external conditions and factors, living and nonliving, that affect an organism or other specified system during its lifetime; the earth's life-support systems for us and for all other forms of life-another term for solar capita. |
| Natural selection | Process by which a particular beneficial gene is reproduced more than other genes in succeeding generations is natural selection. The result of natural selection is a population that contains a greater proportion of organisms better adapted to certain environments. |
| Habitat | The place where an organism lives is called habitat. |
| System | A set of components that function and interact in some regular and theoretically predictable manner is called a system. |
| Input | Input refers to matter, energy, or information entering a system. Compare output, throughput. |
| Echolocation | The use of reflected sound to detect environmental objects. Cetaceans use echolocation to detect prey and avoid obstacles. |
| Predation | Predation refers to a situation in which an organism of one species captures and feeds on parts or all of an organism of another species. |
| Response | The amount of health damage caused by exposure to a certain dose of a harmful substance or form of radiation is a response. |
| Predator | Predator refers to an organism that captures and feeds on parts or all of an organism of another species. |
| Wind | Wind refers to the mass movement of air. |
| Matter | Matter refers to anything that has mass and takes up space. On earth, where gravity is present, we weigh an object to determine its mass. |
| Vertebrates | Vertebrates refer to animals with backbones. Compare with invertebrates. |
| Invertebrates | Invertebrates refers to animals that have no backbones. Compare vertebrates. |
| Base | A substance that combines with a hydrogen ion in solution is called the base. |
| Head | A comparatively high promontory with either a CLIFF or steep face. It extends into a large body of water, such as a sea or lake. An unnamed HEAD is usually called a headland. The section of RIP CURRENT which has widened out seaward of the BREAKERS, also called head of |
| Invertebrate | Animal lacking a backbone is the invertebrate. |
| Conduction | The transfer of heat usually in solids whereby energy is passed from particle to particle by thermal agitation is referred to as conduction. |
| Output | Output refers to matter, energy, or information leaving a system. Compare with input, throughput. |
| Ventral | Ventral refers to of or pertaining to the underside of animals. |
| Nucleus | Extremely tiny center of an atom, making up most of the atom's mass is the nucleus. It contains one or more positively charged protons and one or more neutrons with no electrical charge. |

## Chapter 6. Physiological Analysis Nerve Cells and Behavior

| | |
|---|---|
| Cell | Smallest living unit of an organism. Each cell is encased in an outer membrane or wall and contains genetic material and other parts to perform its life function. Organisms such as bacteria consist of only one cell, but most of the organisms we are. |
| Well | Well refers to a hole, generally cylindrical and usually walled or lined with pipe, that is dug or drilled into the ground to penetrate an aquifer below the zone of saturation. |
| Ion | Ion refers to atom or group of atoms with one or more positive or negative electrical charges. Compare atone, molecule. |
| Atoms | Atoms refers to minute units made of subatomic particles that are the basic building blocks of all chemical elements and thus all matter; the smallest unit of an element that can exist and still have the unique characteristics of that element. Compare to ion, molecule. |
| Energy | Capacity to do work by performing mechanical, physical, chemical, or electrical tasks or to cause a heat transfer between two objects at different temperatures is an energy. |
| Pore | Pore refers to an opening or void space in; oil or rock. |
| Chemical | One of the millions of different elements and compounds found naturally or synthesized by human is referred to as chemical. |
| Concentration | Amount of a chemical in a particular volume or weight of air, water, soil, or other medium is referred to as concentration. |
| Gradient | A measure of slope in meters of rise or fall per meter of horizontal distance. More general, a change of a value per unit of distance, e.g. the GRADIENT in longshore transport causes EROSION or ACCRETION. With reference to winds or currents, the rate of increase or decrease in speed, usually in the vertical; or the curve that represents this rate. |
| State | State refers to an expression of the internal form of matter. Water exists in three states: solid, liquid, and gas. A solid has a fixed volume and fixed shape; a liquid has a fixed volume but no fixed shape; and a gas has neither fixed volume nor fixed shape. |
| Impermeable | Impermeable refers to impervious; the condition of rock that does not allow fluids to flow through it. |
| Point | Point refers to the extreme end of a cape, or the outer end of any land area protruding into the water, usually less prominent than a cape. A low profile shoreline promontory of more or less triangular shape, the top of which extends seaward. |
| Event | Event refers to an occurrence meeting specified conditions, e.g. damage, a threshold wave height or a threshold water level. |
| Wave | An oscillatory movement in a body of water manifested by an alternate rise and fall of the surface is called a wave. Disturbances of the surface of a liquid body, as the ocean, in the form of a ridge, swell or hump. The term wave by itself usually refers to the term surface gravity wave. |
| Threshold | Threshold refers to a point in the operation of a system at which a change occurs. With respect to toxicology, it is a level below which effects are not observable and above which effects become apparent. |
| Situation | The relative geographic location of a site that makes it a good location for a city is a situation. |
| Fire | The rapid combination of oxygen with organic material to produce flame, heat, and light is referred to as the fire. |
| Magnitude | An assessment of the size of an event is a magnitude. Magnitude scales exist for earthquakes, volcanic eruptions, hurricanes, and tornadoes. For earthquakes, different magnitudes are |

## Chapter 6. Physiological Analysis Nerve Cells and Behavior

| | |
|---|---|
| | calculated for the same earthquake when different types of seismic waves are used. |
| Heat | Total kinetic energy of all the randomly moving atoms, ions, or molecules within a given substance, excluding the overall motion of the whole object. This form of kinetic energy flows from one body to another when there is a temperature difference betwe is referred to as heat. |
| Reach | Reach refers to an arm of the ocean extending into the land. A straight section of restricted waterway of considerable extent; may be similar to a narrows, except much longer in extent. |
| Mutation | A random change in DNA molecules making up genes that can yield changes in anatomy, physiology, or behavior in offspring is a mutation. |
| Gene | Gene refers to the fundamental unit in inheritance; it carries the characteristics of parents to their offspring. |
| Protein | Protein refers to a family of complex organic compounds containing nitrogen and composed of various amino acids. |
| Sea | Sea refers to the ocean. A large body of salt water, second in rank to an ocean, more or less landlocked and generally part of, or connected with, an ocean or a larger sea. State of the ocean or lake surface, in regard to waves. |
| Flood | Period when tide level is rising; often taken to mean the flood current which occurs during this period. A flow above the CARRYING CAPACITY of a CHANNEL. |
| Ocean | The great body of salt water which occupies two-thirds of the surface of the Earth, or one of its major subdivisions is called an ocean. |
| Dorsal | The upper or back surface of an animal is a dorsal. |
| Mantle | Zone of the earth's interior between its core and its crust. Compare with core, crust is the mantle. |
| Organism | Any form of life is referred to as an organism. |
| Current | Current refers to the flowing of water, or other liquid or gas. That portion of a stream of water which is moving with a velocity much greater than the average or in which the progress of the water is principally concentrated. Ocean currents can be classified in a number of different ways. |
| Messenger | A free-falling weight that is attached to a cable and used to trigger the closure of water-sampling bottles is referred to as a messenger. |
| Genes | Genes refers to coded units of information about specific traits that are passed on from parents to offspring during reproduction. They consist of segments of DNA molecules found in chromosomes. |
| Lead | Lead refers to a heavy metal that is an important constituent of automobile batteries and other industrial products. A toxic metal capable of causing environmental disruption and producing a health problem to people and other living organisms. |
| Adhesion | Adhesion refers to an attachment of water molecules to other substances by hydrogen bonds. |
| Development | Development refers to change from a society that is largely rural, agricultural, illiterate, and poor, with a rapidly growing population, to one that is mostly urban, industrial, educated, and wealthy, with a slowly growing or stationary population. |
| Recent | A synonym of Holocene is called recent. |
| Host | Plant or animal on which a parasite feeds is referred to as host. |
| Rock | Any material that makes up a large, natural, continuous part of earth's crust is rock. |

# Chapter 6. Physiological Analysis Nerve Cells and Behavior

## Chapter 6. Physiological Analysis Nerve Cells and Behavior

| | |
|---|---|
| Surface water | Precipitation that does not infiltrate the ground or return to the atmosphere by evaporation or transpiration is called surface water. |
| Site | A factor considering the summation of all environmental features of a location that influences the placement of a city is a site. |
| Frequency | Number of events in a given time interval. For earthquakes, it is the number of cycles of seismic waves that pass in a second; frequency = l/period. |
| Glass | Glass refers to matter created when magma cools too quickly for atoms to arrange themselves into the ordered atomic structures of minerals. Most glasses are supercooled liquids. |
| Experiment | Experiment refers to procedure a scientist uses to study some phenomenon under known conditions. Some experiments are conducted in the laboratory, but others are conducted in nature. The resulting scientific data or facts must be verified or confirmed by repeated observation. |
| Reduce | With respect to waste management, reduce refers to practices that will reduce the amount of waste we produce. |
| Efficiency | The ratio of output to input. With machines, usually the ratio of work or power produced to the energy or power used to operate or fuel them. With living things, efficiency may be defined as either the useful work done or the energy stored in a useful fo. |
| Range | Land used for grazing is referred to as the range. |
| Key | Key refers to a low, insular BANK of sand, coral, etc., as one of the islets off the southern coast of Florida. |
| Wavelength | The horizontal distance between corresponding points on successive waves, such as from crest to crest or trough to trough is called a wavelength. |
| Acceleration | Acceleration refers to cause to move faster. The rate of change of motion. |
| Species | Group of organisms that resemble one another in appearance, behavior, chemical makeup and processes, and genetic structure is a species. Organisms that reproduce sexually are classified as members of the same species only if they can breed with one another and produce offspring. |
| Aquatic | Aquatic refers to pertaining to water. |
| Front | The boundary between two air masses with different temperatures and densitie is referred to as front. |
| Canal | Canal refers to an artificial watercourse cut through a land area for such uses as navigation and irrigation. |
| Interference | Interference refers to during succession, one species prevents the entrance of later successional species into an ecosystem. For example, some grasses produce such dense and thick mats that seeds of trees cannot reach the soil to germinate. As long as these grasses persist, th. |
| Observations | Information obtained through one or more of the five senses or through instruments that extend the senses are observations. |
| Tongue | A long narrow strip of land, projecting into a body of water is a tongue. |
| Compression | A state of stress that causes a pushing together or contraction is called compression. |
| Trough | A long and broad submarine depression with gently sloping sides is referred to as a trough. |
| Fall | Fall refers to a mass moving nearly vertical and downward under the influence of gravity. |

## Chapter 6. Physiological Analysis Nerve Cells and Behavior

## Chapter 6. Physiological Analysis Nerve Cells and Behavior

| | |
|---|---|
| Radar | An instrument for determining the distance and direction to an object by measuring the time needed for radio signals to travel from the instrument to the object and back, and by measuring the angle through which the instrument's antenna has traveled is referred to as radar. |
| Altitude | Altitude refers to the height above sea level. Compare to latitude. |
| Accuracy | Accuracy refers to the extent to which a measurement agrees with the accepted or correct value for that quantity, based on careful measurements by many people over a long time. Compare to precision. |
| Precision | Precision refers to a measure of reproducibility, or how closely a series of measurements of the same quantity agree with one another. Compare with accuracy. |
| Tests | The skeleton or shells of certain microorganisms are called tests. |
| Elevation | Elevation refers to the distance of a point above a specified surface of constant potential; the distance is measured along the direction of gravity between the point and the surface. |
| Map | Map refers to a representation of Earth's surface usually depicting mostly land areas. |
| Degree | An arbitrary measure of temperature. One degree Celsius _ 1.8 degrees Fahrenheit. |
| Network | Network refers to a set consisting of stations for which geometric relationships have been determined and which are so related that removal of one station from the set will affect the relationships between the other stations; and lines connecting the stations to show this interdependence. |
| Feedback | A kind of system response that occurs when output of the system also serves as input leading to changes in the system is called feedback. |
| Mass | The amount of material in an object is the mass. |
| Mist | Water vapor suspended in the air in very small drops finer than rain, larger than fog is referred to as mist. |
| Forest | Forest refers to biome with enough average annual precipitation to support growth of various species of trees and smaller forms of vegetation. Compare desert, grassland. |
| Desert | Desert refers to biome in which evaporation exceeds precipitation and the average amount of precipitation is less than 25 centimeters a year. Such areas have little vegetation or have widely spaced, mostly low vegetation. Compare forest, grassland. |
| Stream | Stream refers to any flow of water; a current. A course of water flowing along a bed in the earth. |
| Strand | Strand refers to shore or beach of the ocean or a large lake. The land bordering any large body of water, especially a sea or an arm of the ocean. |
| Depression | Depression refers to a general term signifying any depressed or lower area in the ocean floor. |
| Monitoring | Monitoring refers to the process of collecting data on a regular basis at specific sites to provide a database from which to evaluate change. |
| Conditions | Conditions refers to physical or chemical attributes of the environment that, while not being consumed, influence biological processes and population growth. Examples are temperature, salinity, and acidity. Compare resources. |
| Compound | Compound refers to combination of atoms, or oppositely charged ions, of two or more different elements held together by attractive forces called chemical bonds. Compare element. |

## Chapter 7. Physiological Analysis of Behavior - The Endocrine System

| | |
|---|---|
| System | A set of components that function and interact in some regular and theoretically predictable manner is called a system. |
| Development | Development refers to change from a society that is largely rural, agricultural, illiterate, and poor, with a rapidly growing population, to one that is mostly urban, industrial, educated, and wealthy, with a slowly growing or stationary population. |
| Point | Point refers to the extreme end of a cape, or the outer end of any land area protruding into the water, usually less prominent than a cape. A low profile shoreline promontory of more or less triangular shape, the top of which extends seaward. |
| Chemical | One of the millions of different elements and compounds found naturally or synthesized by human is referred to as chemical. |
| Morphology | Morphology refers to river/estuary/lake/seabed form and its change with time. |
| Environment | Environment refers to all external conditions and factors, living and nonliving, that affect an organism or other specified system during its lifetime; the earth's life-support systems for us and for all other forms of life-another term for solar capita. |
| Host | Plant or animal on which a parasite feeds is referred to as host. |
| Species | Group of organisms that resemble one another in appearance, behavior, chemical makeup and processes, and genetic structure is a species. Organisms that reproduce sexually are classified as members of the same species only if they can breed with one another and produce offspring. |
| Network | Network refers to a set consisting of stations for which geometric relationships have been determined and which are so related that removal of one station from the set will affect the relationships between the other stations; and lines connecting the stations to show this interdependence. |
| Cell | Smallest living unit of an organism. Each cell is encased in an outer membrane or wall and contains genetic material and other parts to perform its life function. Organisms such as bacteria consist of only one cell, but most of the organisms we are. |
| Invertebrates | Invertebrates refers to animals that have no backbones. Compare vertebrates. |
| Response | The amount of health damage caused by exposure to a certain dose of a harmful substance or form of radiation is a response. |
| Duration | In forecasting waves, the length of time the wind blows in essentially the same direction over the FETCH is a duration. |
| Genes | Genes refers to coded units of information about specific traits that are passed on from parents to offspring during reproduction. They consist of segments of DNA molecules found in chromosomes. |
| Concentration | Amount of a chemical in a particular volume or weight of air, water, soil, or other medium is referred to as concentration. |
| Messenger | A free-falling weight that is attached to a cable and used to trigger the closure of water-sampling bottles is referred to as a messenger. |
| Ion | Ion refers to atom or group of atoms with one or more positive or negative electrical charges. Compare atone, molecule. |
| Acid | A substance that releases a hydrogen ion in solution is an acid. |
| Protein | Protein refers to a family of complex organic compounds containing nitrogen and composed of various amino acids. |

# Chapter 7. Physiological Analysis of Behavior - The Endocrine System

## Chapter 7. Physiological Analysis of Behavior - The Endocrine System

| | |
|---|---|
| Vertebrates | Vertebrates refer to animals with backbones. Compare with invertebrates. |
| Solubility | A measure of how readily one chemical substance can dissolve another is a solubility. |
| Nucleus | Extremely tiny center of an atom, making up most of the atom's mass is the nucleus. It contains one or more positively charged protons and one or more neutrons with no electrical charge. |
| Dna | Large molecules in the cells of organisms that carry genetic information in living organisms are dna. |
| Gene | Gene refers to the fundamental unit in inheritance; it carries the characteristics of parents to their offspring. |
| Output | Output refers to matter, energy, or information leaving a system. Compare with input, throughput. |
| Degree | An arbitrary measure of temperature. One degree Celsius _ 1.8 degrees Fahrenheit. |
| Well | Well refers to a hole, generally cylindrical and usually walled or lined with pipe, that is dug or drilled into the ground to penetrate an aquifer below the zone of saturation. |
| Sea | Sea refers to the ocean. A large body of salt water, second in rank to an ocean, more or less landlocked and generally part of, or connected with, an ocean or a larger sea. State of the ocean or lake surface, in regard to waves. |
| Spring | A place where groundwater flows out onto the surface is a spring. |
| Migration | A term that refers to the habit of some animals is a migration. |
| Salinity | Salinity refers to number of grams of salt per thousand grams of seawater, usually expressed in parts per thousand. |
| Metabolism | Ability of a living cell or organism to capture and transform matter and energy from its environment to supply its needs for survival, growth, and reproduction is called metabolism. |
| Fold | Fold refers to wavy geologic structures formed by the compression and bending of sedimentary layers. |
| Exoskeleton | Exoskeleton refers to a skeleton partially or completely covering the exterior of a plant or animal. |
| Larva | The immature form of an animal that differs significantly from the adult form is a larva. |
| Head | A comparatively high promontory with either a CLIFF or steep face. It extends into a large body of water, such as a sea or lake. An unnamed HEAD is usually called a headland. The section of RIP CURRENT which has widened out seaward of the BREAKERS, also called head of |
| Molt | The process of dispensing with an exoskeleton in order to secrete a bigger exoskeleton that will accommodate a larger body size is referred to as molt. |
| Temperature | Temperature refers to a measure of the average speed of motion of the atoms, ions, or molecules in a substance or combination of substances at a given moment. Compare with heat. |
| Dorsal | The upper or back surface of an animal is a dorsal. |
| Front | The boundary between two air masses with different temperatures and densitie is referred to as front. |
| Fall | Fall refers to a mass moving nearly vertical and downward under the influence of gravity. |
| Positive feedback | Positive feedback refers to a type of feedback that occurs when an increase in output leads to a further increase in output. This is sometimes known as a vicious cycle, since the more you have the more you get. |

## Chapter 7. Physiological Analysis of Behavior - The Endocrine System

| | |
|---|---|
| Feedback | A kind of system response that occurs when output of the system also serves as input leading to changes in the system is called feedback. |
| Surge | Surge refers to long-interval variations in velocity and pressure in fluid flow, not necessarily periodic, perhaps even transient in nature. The name applied to wave motion with a period intermediate between that of an ordinary wind wave and that of the tide. Changes in water level as a result of meteorological forcing causing a difference between the recorded water level and that predicted using harmonic analysis, may be positive or negative. |
| Positive feedback loop | Positive feedback loop refers to a situation in which a change in a certain direction provides information that causes a system to change further in the same direction. This can lead to a runaway or vicious cycle. Compare with negative feedback loop. |
| Feedback loop | Feedback loop refers to circuit of sensing, evaluating, and reacting to changes in environmental conditions as a result of information fed back into a system; it occurs when one change leads to some other change, which then eventually either reinforces or slows the original cha. |
| Loop | That part of a STANDING WAVE where the vertical motion is greatest and the horizontal velocities are least is a loop. |
| Substrate | A general term in reference to the surface on or within which organisms live is a substrate. |
| Lead | Lead refers to a heavy metal that is an important constituent of automobile batteries and other industrial products. A toxic metal capable of causing environmental disruption and producing a health problem to people and other living organisms. |
| Hierarchy | Hierarchy refers to grouping of objects by degrees of complexity, grade, or class. A hierarchical system of nomenclature is based on distinctions within groups and between groups. |
| Tongue | A long narrow strip of land, projecting into a body of water is a tongue. |
| Heat | Total kinetic energy of all the randomly moving atoms, ions, or molecules within a given substance, excluding the overall motion of the whole object. This form of kinetic energy flows from one body to another when there is a temperature difference betwe is referred to as heat. |
| Neck | Neck refers to the narrow strip of land which connects a peninsula with the mainland, or connects two ridges. The narrow band of water flowing seaward through the surf. |
| Radioactivity | Nuclear change in which unstable nuclei of atoms spontaneously shoot out 'chunks' of mass, energy, or both, at a fixed rate is called radioactivity. The three principal types of radioactivity are gamma rays and fast-moving alpha particles and beta particles. |
| Conditions | Conditions refers to physical or chemical attributes of the environment that, while not being consumed, influence biological processes and population growth. Examples are temperature, salinity, and acidity. Compare resources. |
| Mist | Water vapor suspended in the air in very small drops finer than rain, larger than fog is referred to as mist. |
| Site | A factor considering the summation of all environmental features of a location that influences the placement of a city is a site. |
| Recent | A synonym of Holocene is called recent. |
| Reach | Reach refers to an arm of the ocean extending into the land. A straight section of restricted waterway of considerable extent; may be similar to a narrows, except much longer in extent. |
| State | State refers to an expression of the internal form of matter. Water exists in three states: |

## Chapter 7. Physiological Analysis of Behavior - The Endocrine System

| | |
|---|---|
| | solid, liquid, and gas. A solid has a fixed volume and fixed shape; a liquid has a fixed volume but no fixed shape; and a gas has neither fixed volume nor fixed shape. |
| Reproduction | Production of offspring by one is called reproduction. |
| Experiment | Experiment refers to procedure a scientist uses to study some phenomenon under known conditions. Some experiments are conducted in the laboratory, but others are conducted in nature. The resulting scientific data or facts must be verified or confirmed by repeated observation. |
| Situation | The relative geographic location of a site that makes it a good location for a city is a situation. |
| Probability | A mathematical statement about how likely it is that something will happen is probability. |
| Annual | Annual refers to plant that grows, sets seed, and dies in one growing season. Compare to perennial. |
| Zone | Division or province of the ocean with homogeneous characteristics is referred to as a zone. |
| Domesticated species | Wild species tamed or genetically altered by crossbreeding for use by humans for food, pets, or enjoyment are referred to as domesticated species. |
| Current | Current refers to the flowing of water, or other liquid or gas. That portion of a stream of water which is moving with a velocity much greater than the average or in which the progress of the water is principally concentrated. Ocean currents can be classified in a number of different ways. |
| Limestone | A sedimentary rock composed dominantly of calcium carbonate, either precipitated from seawater or deposited as shell debris are called the limestone. |
| Entrance | The entrance to a navigable BAY, HARBOR or CHANNEL, INLET or mouth separating the ocean from an inland water body. |
| Mass | The amount of material in an object is the mass. |
| Disperse | Disperse refers to to spread or distribute from a fixed or constant source. To cause to become widely separated. |
| Matter | Matter refers to anything that has mass and takes up space. On earth, where gravity is present, we weigh an object to determine its mass. |
| Desert | Desert refers to biome in which evaporation exceeds precipitation and the average amount of precipitation is less than 25 centimeters a year. Such areas have little vegetation or have widely spaced, mostly low vegetation. Compare forest, grassland. |
| Range | Land used for grazing is referred to as the range. |
| Frequency | Number of events in a given time interval. For earthquakes, it is the number of cycles of seismic waves that pass in a second; frequency = l/period. |
| Risk | Risk refers to the probability that something undesirable will happen from deliberate or accidental exposure. |
| Benefits | Benefits refers to the economic value of a scheme, usually measured in terms of the cost of damages avoided by the scheme, or the valuation of perceived amenity or environmental improvements. |
| Natural selection | Process by which a particular beneficial gene is reproduced more than other genes in succeeding generations is natural selection. The result of natural selection is a population that contains a greater proportion of organisms better adapted to certain environments. |
| Theory | A general explanation of a characteristic of nature consistently supported by observation or |

## Chapter 7. Physiological Analysis of Behavior - The Endocrine System

| | |
|---|---|
| | experiment is referred to as a theory. |
| Interference | Interference refers to during succession, one species prevents the entrance of later successional species into an ecosystem. For example, some grasses produce such dense and thick mats that seeds of trees cannot reach the soil to germinate. As long as these grasses persist, th. |
| Competition | Two or more individual organisms of a single species or two or more individuals of different species attempting to use the same scarce resources in the same ecosystem is a competition. |
| Stress | Stress refers to force per unit area. May be compression, tension, or shear. |
| Coast | A strip of land of indefinite length and width that extends from the SEASHORE inland to the first major change in terrain features is referred to as coast. |
| Intertidal zone | The area of shoreline between low and high tides are called the intertidal zone. |
| Swim bladder | Swim bladder refers to a gas-filled pouch in many bony fishes that is used to attain neutral buoyancy by regulating the amount of gas in the bladder. |
| Population density | Number of organisms in a particular population found in a specified area is the population density. Compare with ecological population density. |
| Density | Density refers to the ratio of a mass to a unit volume specified as grams per cubic centimeter. |
| Absorption | Conversion of sound or light energy into heat is called absorption. |
| Key | Key refers to a low, insular BANK of sand, coral, etc., as one of the islets off the southern coast of Florida. |
| Food | General term for organic molecules capable of providing energy to heterotrophs when combined with oxygen during biochemical respiration is called food. |
| Observations | Information obtained through one or more of the five senses or through instruments that extend the senses are observations. |

## Chapter 8. The Development of Behavior

| | |
|---|---|
| Development | Development refers to change from a society that is largely rural, agricultural, illiterate, and poor, with a rapidly growing population, to one that is mostly urban, industrial, educated, and wealthy, with a slowly growing or stationary population. |
| System | A set of components that function and interact in some regular and theoretically predictable manner is called a system. |
| State | State refers to an expression of the internal form of matter. Water exists in three states: solid, liquid, and gas. A solid has a fixed volume and fixed shape; a liquid has a fixed volume but no fixed shape; and a gas has neither fixed volume nor fixed shape. |
| Morphology | Morphology refers to river/estuary/lake/seabed form and its change with time. |
| Genes | Genes refers to coded units of information about specific traits that are passed on from parents to offspring during reproduction. They consist of segments of DNA molecules found in chromosomes. |
| Environment | Environment refers to all external conditions and factors, living and nonliving, that affect an organism or other specified system during its lifetime; the earth's life-support systems for us and for all other forms of life-another term for solar capita. |
| Homeostasis | Maintenance of favorable internal conditions in a system despite fluctuations in external condition are called homeostasis. |
| Predator | Predator refers to an organism that captures and feeds on parts or all of an organism of another species. |
| Lake | Large natural body of standing fresh water formed when water from precipitation, land runoff, or groundwater flow fills a depression in the earth created by glaciation, earth movement, volcanic activity, or a giant meteorit are called the lake. |
| Response | The amount of health damage caused by exposure to a certain dose of a harmful substance or form of radiation is a response. |
| Species | Group of organisms that resemble one another in appearance, behavior, chemical makeup and processes, and genetic structure is a species. Organisms that reproduce sexually are classified as members of the same species only if they can breed with one another and produce offspring. |
| Dorsal | The upper or back surface of an animal is a dorsal. |
| Nip | The cut made by waves in a shoreline of emergence is called a nip. |
| Food | General term for organic molecules capable of providing energy to heterotrophs when combined with oxygen during biochemical respiration is called food. |
| Oil | The liquid form of petroleum consisting of a complex mixture of large hydrocarbon molecules is referred to as oil. |
| Head | A comparatively high promontory with either a CLIFF or steep face. It extends into a large body of water, such as a sea or lake. An unnamed HEAD is usually called a headland. The section of RIP CURRENT which has widened out seaward of the BREAKERS, also called head of |
| Glass | Glass refers to matter created when magma cools too quickly for atoms to arrange themselves into the ordered atomic structures of minerals. Most glasses are supercooled liquids. |
| Chemical | One of the millions of different elements and compounds found naturally or synthesized by human is referred to as chemical. |
| Compass | An instrument for showing direction by means of a magnetic needle swinging freely on a pivot and pointing to magnetic north are called compass. |

# Chapter 8. The Development of Behavior

## Chapter 8. The Development of Behavior

| | |
|---|---|
| Conditions | Conditions refers to physical or chemical attributes of the environment that, while not being consumed, influence biological processes and population growth. Examples are temperature, salinity, and acidity. Compare resources. |
| Age structure | Percentage of the population at each age level in a population is referred to as the age structure. |
| Frequency | Number of events in a given time interval. For earthquakes, it is the number of cycles of seismic waves that pass in a second; frequency = l/period. |
| River | A natural stream of water larger than a brook or creek is a river. |
| Zooplankton | Zooplankton refers to animal plankton. Small floating herbivores that feed on plant plankton. Compare with phytoplankton. |
| Filter feeders | Animals that feed by 'sifting' small organisms or organic particles that are suspended in the water are filter feeders. |
| Strain | A change in torn or size of a body due to external forces is called strain. |
| Point | Point refers to the extreme end of a cape, or the outer end of any land area protruding into the water, usually less prominent than a cape. A low profile shoreline promontory of more or less triangular shape, the top of which extends seaward. |
| Prey | Organism that is captured and serves as a source of food for an organism of another species is called prey. |
| Well | Well refers to a hole, generally cylindrical and usually walled or lined with pipe, that is dug or drilled into the ground to penetrate an aquifer below the zone of saturation. |
| Neck | Neck refers to the narrow strip of land which connects a peninsula with the mainland, or connects two ridges. The narrow band of water flowing seaward through the surf. |
| Slide | In mass wasting, movement of a descending mass along a plane approximately parallel to the slope of the surface is referred to as slide. |
| Benefits | Benefits refers to the economic value of a scheme, usually measured in terms of the cost of damages avoided by the scheme, or the valuation of perceived amenity or environmental improvements. |
| Reach | Reach refers to an arm of the ocean extending into the land. A straight section of restricted waterway of considerable extent; may be similar to a narrows, except much longer in extent. |
| Delta | ALLUVIAL DEPOSIT, usually triangular, at the mouth of a river of other stream. It is normally built up only where there is no tidal or CURRENT action capable of removing the sediment as fast as it is deposited, and hence the DELTA builds forward from the COASTLINE. A TIDAL DELTA is a similar deposit at the mouth of a tidal INLET, put there by TIDAL CURRENTS. A WAVE DELTA is a deposit made by large waves which run over the top of a SPIT or BAR beach and down the landward side. |
| Ecology | Ecology refers to study of the interactions of living organisms with one another and with their nonliving environment of matter and energy; study of the structure and functions of nature. |
| Range | Land used for grazing is referred to as the range. |
| Reflection | The process by which the energy of the wave is returned seaward is a reflection. |
| Chromosome | A grouping of various genes and associated proteins in plant and animal cells that carry certain types of genetic informatio are called the chromosome. |
| Duration | In forecasting waves, the length of time the wind blows in essentially the same direction |

## Chapter 8. The Development of Behavior

## Chapter 8. The Development of Behavior

| | |
|---|---|
| | over the FETCH is a duration. |
| Feedback | A kind of system response that occurs when output of the system also serves as input leading to changes in the system is called feedback. |
| Output | Output refers to matter, energy, or information leaving a system. Compare with input, throughput. |
| Peak period | The wave period determined by the inverse of the frequency at which the wave energy spectrum reaches its maximum is called peak period. |
| Marsh | Marsh refers to a tract of soft, wet land, usually vegetated by reeds, grasses and occasionally small shrubs. Soft, wet area periodically or continuously flooded to a shallow depth, usually characterized by a particular subclass of grasses, cattails and other low plants. |
| Accuracy | Accuracy refers to the extent to which a measurement agrees with the accepted or correct value for that quantity, based on careful measurements by many people over a long time. Compare to precision. |
| Network | Network refers to a set consisting of stations for which geometric relationships have been determined and which are so related that removal of one station from the set will affect the relationships between the other stations; and lines connecting the stations to show this interdependence. |
| Population | Group of individual organisms of the same species living within a particular area is referred to as a population. |
| Density | Density refers to the ratio of a mass to a unit volume specified as grams per cubic centimeter. |
| Recent | A synonym of Holocene is called recent. |
| Experiment | Experiment refers to procedure a scientist uses to study some phenomenon under known conditions. Some experiments are conducted in the laboratory, but others are conducted in nature. The resulting scientific data or facts must be verified or confirmed by repeated observation. |
| Observations | Information obtained through one or more of the five senses or through instruments that extend the senses are observations. |
| Hypothesis | In science, an explanation set forth in a manner that can be tested and is capable of being disproved. A tested hypothesis is accepted until and unless it has been disproved. |
| Habitat | The place where an organism lives is called habitat. |
| Stress | Stress refers to force per unit area. May be compression, tension, or shear. |
| Gravel | Gravel refers to loose, rounded fragments of rock, larger than sand, but smaller than cobbles. Small stones and pebbles, or a mixture of these with sand. |
| Sea | Sea refers to the ocean. A large body of salt water, second in rank to an ocean, more or less landlocked and generally part of, or connected with, an ocean or a larger sea. State of the ocean or lake surface, in regard to waves. |
| Stream | Stream refers to any flow of water; a current. A course of water flowing along a bed in the earth. |
| Migration | A term that refers to the habit of some animals is a migration. |
| Competitive exclusion | No two species can occupy exactly the same fundamental niche indefinitely in a habitat where there is not enough of a particular resource to meet the needs of both specie is a |

## Chapter 8. The Development of Behavior

| | |
|---|---|
| | competitive exclusion. |
| Input | Input refers to matter, energy, or information entering a system. Compare output, throughput. |
| Accumulation | Buildup of matter, energy, or information in a system is referred to as accumulation. |
| Reproduction | Production of offspring by one is called reproduction. |
| Site | A factor considering the summation of all environmental features of a location that influences the placement of a city is a site. |
| Situation | The relative geographic location of a site that makes it a good location for a city is a situation. |
| Fat | Fat refers to an organic compound composed of carbon, hydrogen, and oxygen that is insoluble in water. |
| Spring | A place where groundwater flows out onto the surface is a spring. |
| Degree | An arbitrary measure of temperature. One degree Celsius _ 1.8 degrees Fahrenheit. |
| Reduce | With respect to waste management, reduce refers to practices that will reduce the amount of waste we produce. |
| Front | The boundary between two air masses with different temperatures and densitie is referred to as front. |
| Current | Current refers to the flowing of water, or other liquid or gas. That portion of a stream of water which is moving with a velocity much greater than the average or in which the progress of the water is principally concentrated. Ocean currents can be classified in a number of different ways. |
| Surge | Surge refers to long-interval variations in velocity and pressure in fluid flow, not necessarily periodic, perhaps even transient in nature. The name applied to wave motion with a period intermediate between that of an ordinary wind wave and that of the tide. Changes in water level as a result of meteorological forcing causing a difference between the recorded water level and that predicted using harmonic analysis, may be positive or negative. |
| Threshold | Threshold refers to a point in the operation of a system at which a change occurs. With respect to toxicology, it is a level below which effects are not observable and above which effects become apparent. |
| Buffer | A group of substances that tend to work against change in the pH of a solution by combining with free ions is called buffer. |
| Stability | Ability of a living system to withstand or recover from externally imposed changes or stresses is called stability. |
| Resilience | Ability of a living system to restore itself to the original condition after being exposed to an outside disturbance that is not too drastic is referred to as resilience. |
| Host | Plant or animal on which a parasite feeds is referred to as host. |
| Matter | Matter refers to anything that has mass and takes up space. On earth, where gravity is present, we weigh an object to determine its mass. |
| Lead | Lead refers to a heavy metal that is an important constituent of automobile batteries and other industrial products. A toxic metal capable of causing environmental disruption and producing a health problem to people and other living organisms. |
| Community | Community refers to populations of all species living and interacting in an area at a particular time. |

# Chapter 8. The Development of Behavior

## Chapter 8. The Development of Behavior

| | |
|---|---|
| **Competition** | Two or more individual organisms of a single species or two or more individuals of different species attempting to use the same scarce resources in the same ecosystem is a competition. |

# Chapter 8. The Development of Behavior

## Chapter 9. Biological Clocks

| | |
|---|---|
| Lunar day | The time of rotation of the Earth with respect to the moon, or the interval between two successive upper transits of the moon over the meridian of a place is a lunar day. The mean lunar day is approximately 24.84 solar hours in length, or 1.035 times as great as the mean solar day. |
| Annual | Annual refers to plant that grows, sets seed, and dies in one growing season. Compare to perennial. |
| Conditions | Conditions refers to physical or chemical attributes of the environment that, while not being consumed, influence biological processes and population growth. Examples are temperature, salinity, and acidity. Compare resources. |
| Stability | Ability of a living system to withstand or recover from externally imposed changes or stresses is called stability. |
| Entrainment | The act of resuspending or remobilizing a sediment grain resting on the sea bottom is an entrainment. |
| Temperature | Temperature refers to a measure of the average speed of motion of the atoms, ions, or molecules in a substance or combination of substances at a given moment. Compare with heat. |
| Event | Event refers to an occurrence meeting specified conditions, e.g. damage, a threshold wave height or a threshold water level. |
| Hierarchy | Hierarchy refers to grouping of objects by degrees of complexity, grade, or class. A hierarchical system of nomenclature is based on distinctions within groups and between groups. |
| Food | General term for organic molecules capable of providing energy to heterotrophs when combined with oxygen during biochemical respiration is called food. |
| Chart | A map that depicts mostly water and the adjoining land areas is referred to as chart. |
| Organism | Any form of life is referred to as an organism. |
| Well | Well refers to a hole, generally cylindrical and usually walled or lined with pipe, that is dug or drilled into the ground to penetrate an aquifer below the zone of saturation. |
| Cyanobacteria | Cyanobacteria refers to single-celled, prokaryotic, microscopic organisms. Before being reclassified as monera, they were called blue-green algae. |
| Relative humidity | Relative humidity refers to the amount of water vapor in a certain mass of air, expressed as a percentage of the maximum amount it could hold at that temperature. Compare with absolute humidity. |
| Humidity | A measure of the amount of water vapor in an air mass is a humidity. |
| Radiation | Fast-moving particles or waves of energy are called radiation. |
| High tide | The high-water position corresponding to a tidal crest are called the high tide. |
| Tide | The periodic rising and falling of the water that results from gravitational attraction of the moon and sun acting upon the rotating earth is the tide. Although the accompanying horizontal movement of the water resulting from the same cause is also sometimes called the tide, it is preferable to designate the latter as tidal current, reserving the name tide for the vertical movement. |
| Environment | Environment refers to all external conditions and factors, living and nonliving, that affect an organism or other specified system during its lifetime; the earth's life-support systems for us and for all other forms of life-another term for solar capita. |
| Intertidal | The zone between the high and LOW WATER marks is an intertidal. |

## Chapter 9. Biological Clocks

## Chapter 9. Biological Clocks

| | |
|---|---|
| Desiccation | Desiccation refers to drying. |
| Spring | A place where groundwater flows out onto the surface is a spring. |
| Tides | The periodic rise and fall of the Earth's water surface as a consequence of the gravitational attraction of the Moon and the Sun, which are called tides. |
| Magnetic field | A region where magnetic forces affect any magnetized bodies or electric currents is the magnetic field. Earth is surrounded by a magnetic field. |
| State | State refers to an expression of the internal form of matter. Water exists in three states: solid, liquid, and gas. A solid has a fixed volume and fixed shape; a liquid has a fixed volume but no fixed shape; and a gas has neither fixed volume nor fixed shape. |
| Point | Point refers to the extreme end of a cape, or the outer end of any land area protruding into the water, usually less prominent than a cape. A low profile shoreline promontory of more or less triangular shape, the top of which extends seaward. |
| Species | Group of organisms that resemble one another in appearance, behavior, chemical makeup and processes, and genetic structure is a species. Organisms that reproduce sexually are classified as members of the same species only if they can breed with one another and produce offspring. |
| Tests | The skeleton or shells of certain microorganisms are called tests. |
| Energy | Capacity to do work by performing mechanical, physical, chemical, or electrical tasks or to cause a heat transfer between two objects at different temperatures is an energy. |
| Bar | An offshore ridge or mound of sand, gravel, or other unconsolidated material which is submerged, especially at the mouth of a river or estuary, or lying parallel to, and a short distance from, the beach is referred to as a bar. |
| Ocean | The great body of salt water which occupies two-thirds of the surface of the Earth, or one of its major subdivisions is called an ocean. |
| Reach | Reach refers to an arm of the ocean extending into the land. A straight section of restricted waterway of considerable extent; may be similar to a narrows, except much longer in extent. |
| Shoreline | Shoreline refers to the intersection of a specified plane of water with the shore. All of the water areas of the state, including reservoirs and their associated uplands, together with the lands underlying them. |
| Seashore | Seashore refers to all ground between the ordinary high-water and low-water mark. The shore of the sea or ocean. |
| Intertidal zone | The area of shoreline between low and high tides are called the intertidal zone. |
| Zone | Division or province of the ocean with homogeneous characteristics is referred to as a zone. |
| Marsh | Marsh refers to a tract of soft, wet land, usually vegetated by reeds, grasses and occasionally small shrubs. Soft, wet area periodically or continuously flooded to a shallow depth, usually characterized by a particular subclass of grasses, cattails and other low plants. |
| Sea | Sea refers to the ocean. A large body of salt water, second in rank to an ocean, more or less landlocked and generally part of, or connected with, an ocean or a larger sea. State of the ocean or lake surface, in regard to waves. |
| Habitat | The place where an organism lives is called habitat. |
| Surf | Collective term for breakers is surf. The wave activity in the area between the shoreline and the outermost limit of breakers. The term surf in literature usually refers to the breaking |

## Chapter 9. Biological Clocks

## Chapter 9. Biological Clocks

| | |
|---|---|
| | waves on shore and on reefs when accompanied by a roaring noise caused by the larger waves breaking. |
| Sand | Sand refers to an unconsolidated mixture of inorganic soil consisting of small but easily distinguishable grains ranging in size from about .062 mm to 2.0 mm. |
| Offshore | Offshore in beach terminology refers to the comparatively flat zone of variable width, extending from the shoreface to the edge of the continental shelf. It is continually submerged. The direction seaward from the shore. The zone beyond the nearshore zone where sediment motion induced by waves alone effectively ceases and where the influence of the sea bed on wave action is small in comparison with the effect of wind. The breaker zone directly seaward of the low tide line. |
| Head | A comparatively high promontory with either a CLIFF or steep face. It extends into a large body of water, such as a sea or lake. An unnamed HEAD is usually called a headland. The section of RIP CURRENT which has widened out seaward of the BREAKERS, also called head of |
| Wave | An oscillatory movement in a body of water manifested by an alternate rise and fall of the surface is called a wave. Disturbances of the surface of a liquid body, as the ocean, in the form of a ridge, swell or hump. The term wave by itself usually refers to the term surface gravity wave. |
| Spring tide | A tide that occurs at or near the time of new or full moon, and which rises highest and falls lowest from the mean sea level is called the spring tide. |
| Storm | Local or regional atmospheric disturbance characterized by strong winds often accompanied by precipitation is referred to as a storm. |
| Population | Group of individual organisms of the same species living within a particular area is referred to as a population. |
| Amplitude | Half of the peak-to-trough range of a wave is the amplitude. |
| Development | Development refers to change from a society that is largely rural, agricultural, illiterate, and poor, with a rapidly growing population, to one that is mostly urban, industrial, educated, and wealthy, with a slowly growing or stationary population. |
| Polychaete | A class of annelid worms that includes most marine segmented worms is called polychaete. |
| Reproduction | Production of offspring by one is called reproduction. |
| Probability | A mathematical statement about how likely it is that something will happen is probability. |
| Coral | Coral refers to any of more than 6,000 species of small cnidarians, many of which are capable of generating hard calcareous skeletons. |
| Base | A substance that combines with a hydrogen ion in solution is called the base. |
| Slide | In mass wasting, movement of a descending mass along a plane approximately parallel to the slope of the surface is referred to as slide. |
| Response | The amount of health damage caused by exposure to a certain dose of a harmful substance or form of radiation is a response. |
| Plants | Eukaryotic, mostly multicelled organisms such as algae, mosses, ferns, flowers, cacti, grasses, beans, wheat, rice, and trees are plants. These organisms use photosynthesis to produce organic nutrients for themselves and for other organisms is referred to as plants. |
| Weather | Weather refers to short-term changes in the temperature, barometric pressure, humidity, precipitation, sunshine, cloud cover, wind direction and speed, and other conditions in the troposphere at a given place and time. Compare with climate. |

# Chapter 9. Biological Clocks

## Chapter 9. Biological Clocks

| | |
|---|---|
| Migration | A term that refers to the habit of some animals is a migration. |
| Mass | The amount of material in an object is the mass. |
| Fuel | Air,/ substar ce that produces heat by combustion is a fuel. |
| Reduce | With respect to waste management, reduce refers to practices that will reduce the amount of waste we produce. |
| Molt | The process of dispensing with an exoskeleton in order to secrete a bigger exoskeleton that will accommodate a larger body size is referred to as molt. |
| Chemical | One of the millions of different elements and compounds found naturally or synthesized by human is referred to as chemical. |
| Constancy | Ability of a living system, such as a population, to maintain a certain size. Compare inertia, resilienc is called constancy. |
| Reflection | The process by which the energy of the wave is returned seaward is a reflection. |
| Precision | Precision refers to a measure of reproducibility, or how closely a series of measurements of the same quantity agree with one another. Compare with accuracy. |
| Cape | Cape refers to a relatively extensive land area jutting seaward from a continent or large island which prominently marks a change in, or interrupts notably, the coastal trend; a prominent feature. |
| Fall | Fall refers to a mass moving nearly vertical and downward under the influence of gravity. |
| Range | Land used for grazing is referred to as the range. |
| Atmosphere | Atmosphere refers to the whole mass of air surrounding the earth. |
| Reference point | Reference point refers to a specified location to which measurements are referred. In beach material studies, a specified point within the reference zone. |
| Compass | An instrument for showing direction by means of a magnetic needle swinging freely on a pivot and pointing to magnetic north are called compass. |
| Desert | Desert refers to biome in which evaporation exceeds precipitation and the average amount of precipitation is less than 25 centimeters a year. Such areas have little vegetation or have widely spaced, mostly low vegetation. Compare forest, grassland. |
| System | A set of components that function and interact in some regular and theoretically predictable manner is called a system. |
| Cell | Smallest living unit of an organism. Each cell is encased in an outer membrane or wall and contains genetic material and other parts to perform its life function. Organisms such as bacteria consist of only one cell, but most of the organisms we are. |
| Genes | Genes refers to coded units of information about specific traits that are passed on from parents to offspring during reproduction. They consist of segments of DNA molecules found in chromosomes. |
| Gene | Gene refers to the fundamental unit in inheritance; it carries the characteristics of parents to their offspring. |
| Chromosome | A grouping of various genes and associated proteins in plant and animal cells that carry certain types of genetic informatio are called the chromosome. |
| Input | Input refers to matter, energy, or information entering a system. Compare output, throughput. |
| Compound | Compound refers to combination of atoms, or oppositely charged ions, of two or more different elements held together by attractive forces called chemical bonds. Compare element. |

## Chapter 9. Biological Clocks

| | |
|---|---|
| Invertebrates | Invertebrates refers to animals that have no backbones. Compare vertebrates. |
| Vertebrates | Vertebrates refer to animals with backbones. Compare with invertebrates. |
| Site | A factor considering the summation of all environmental features of a location that influences the placement of a city is a site. |
| Output | Output refers to matter, energy, or information leaving a system. Compare with input, throughput. |
| Experiment | Experiment refers to procedure a scientist uses to study some phenomenon under known conditions. Some experiments are conducted in the laboratory, but others are conducted in nature. The resulting scientific data or facts must be verified or confirmed by repeated observation. |
| Magnitude | An assessment of the size of an event is a magnitude. Magnitude scales exist for earthquakes, volcanic eruptions. hurricanes, and tornadoes. For earthquakes, different magnitudes are calculated for the same earthquake when different types of seismic waves are used. |
| Host | Plant or animal on which a parasite feeds is referred to as host. |
| Oil | The liquid form of petroleum consisting of a complex mixture of large hydrocarbon molecules is referred to as oil. |
| Nucleus | Extremely tiny center of an atom, making up most of the atom's mass is the nucleus. It contains one or more positively charged protons and one or more neutrons with no electrical charge. |
| Observations | Information obtained through one or more of the five senses or through instruments that extend the senses are observations. |
| Hypothesis | In science, an explanation set forth in a manner that can be tested and is capable of being disproved. A tested hypothesis is accepted until and unless it has been disproved. |
| Core | Core refers to a cylindrical sample extracted from a beach or seabed to investigate the types and DEPTHS of sediment layers. An inner, often much less permeable portion of a BREAKWATER, or BARRIER beach. |
| Nutrients | Nutrients refer to chemicals such as phosphorus and nitrogen that, when released into water sources, may cause pollution events such as eutrophication. |
| Mutation | A random change in DNA molecules making up genes that can yield changes in anatomy, physiology, or behavior in offspring is a mutation. |
| Recent | A synonym of Holocene is called recent. |
| Feedback loop | Feedback loop refers to circuit of sensing, evaluating, and reacting to changes in environmental conditions as a result of information fed back into a system; it occurs when one change leads to some other change, which then eventually either reinforces or slows the original cha. |
| Loop | That part of a STANDING WAVE where the vertical motion is greatest and the horizontal velocities are least is a loop. |
| Dna | Large molecules in the cells of organisms that carry genetic information in living organisms are dna. |
| Protein | Protein refers to a family of complex organic compounds containing nitrogen and composed of various amino acids. |
| Time delay | Time lag between the input of a stimulus into a system and the response to the stimulus is called a time delay. |

# Chapter 9. Biological Clocks

## Chapter 9. Biological Clocks

| | |
|---|---|
| **Phosphate** | An essential plant macronutrient is phosphate. |
| **Accumulation** | Buildup of matter, energy, or information in a system is referred to as accumulation. |
| **Degradation** | The geologic process by means of which various parts of the surface of the earth are worn away and their general level lowered, by the action of wind and water is referred to as degradation. |

## Chapter 10. Mechanisms of Orientation

| | |
|---|---|
| Star | A massive sphere of incandescent gases powered by the conversion of hydrogen to helium and other heavier elements is a star. |
| Compass | An instrument for showing direction by means of a magnetic needle swinging freely on a pivot and pointing to magnetic north are called compass. |
| Magnetic field | A region where magnetic forces affect any magnetized bodies or electric currents is the magnetic field. Earth is surrounded by a magnetic field. |
| Map | Map refers to a representation of Earth's surface usually depicting mostly land areas. |
| Chemical | One of the millions of different elements and compounds found naturally or synthesized by human is referred to as chemical. |
| Echolocation | The use of reflected sound to detect environmental objects. Cetaceans use echolocation to detect prey and avoid obstacles. |
| Head | A comparatively high promontory with either a CLIFF or steep face. It extends into a large body of water, such as a sea or lake. An unnamed HEAD is usually called a headland. The section of RIP CURRENT which has widened out seaward of the BREAKERS, also called head of |
| Spring | A place where groundwater flows out onto the surface is a spring. |
| River | A natural stream of water larger than a brook or creek is a river. |
| Dam | Dam refers to structure built in rivers or estuaries, basically to separate water at both sides and/or to retain water at one side. |
| Surge | Surge refers to long-interval variations in velocity and pressure in fluid flow, not necessarily periodic, perhaps even transient in nature. The name applied to wave motion with a period intermediate between that of an ordinary wind wave and that of the tide. Changes in water level as a result of meteorological forcing causing a difference between the recorded water level and that predicted using harmonic analysis, may be positive or negative. |
| Environment | Environment refers to all external conditions and factors, living and nonliving, that affect an organism or other specified system during its lifetime; the earth's life-support systems for us and for all other forms of life-another term for solar capita. |
| Migration | A term that refers to the habit of some animals is a migration. |
| Habitat | The place where an organism lives is called habitat. |
| Food | General term for organic molecules capable of providing energy to heterotrophs when combined with oxygen during biochemical respiration is called food. |
| Key | Key refers to a low, insular BANK of sand, coral, etc., as one of the islets off the southern coast of Florida. |
| Species | Group of organisms that resemble one another in appearance, behavior, chemical makeup and processes, and genetic structure is a species. Organisms that reproduce sexually are classified as members of the same species only if they can breed with one another and produce offspring. |
| Response | The amount of health damage caused by exposure to a certain dose of a harmful substance or form of radiation is a response. |
| Humidity | A measure of the amount of water vapor in an air mass is a humidity. |
| Rocks | An aggregate of one or more minerals rather large in area are rocks. The three classes of rocks are the following: Igneous rock - crystalline rocks formed from molten material. Sedimentary rock - A rock resulting from the consolidation of loose sediment that has accumulated in layers. Metamorphic rock - Rock that has formed from preexisting rock as a |

# Chapter 10. Mechanisms of Orientation

## Chapter 10. Mechanisms of Orientation

| | |
|---|---|
| | result of heat or pressure. |
| Conditions | Conditions refers to physical or chemical attributes of the environment that, while not being consumed, influence biological processes and population growth. Examples are temperature, salinity, and acidity. Compare resources. |
| Vertebrates | Vertebrates refer to animals with backbones. Compare with invertebrates. |
| Larva | The immature form of an animal that differs significantly from the adult form is a larva. |
| Stream | Stream refers to any flow of water; a current. A course of water flowing along a bed in the earth. |
| Lake | Large natural body of standing fresh water formed when water from precipitation, land runoff, or groundwater flow fills a depression in the earth created by glaciation, earth movement, volcanic activity, or a giant meteorit are called the lake. |
| Mud | Mud refers to a mixture of silt and clay sized particles. |
| Experiment | Experiment refers to procedure a scientist uses to study some phenomenon under known conditions. Some experiments are conducted in the laboratory, but others are conducted in nature. The resulting scientific data or facts must be verified or confirmed by repeated observation. |
| Gradient | A measure of slope in meters of rise or fall per meter of horizontal distance. More general, a change of a value per unit of distance, e.g. the GRADIENT in longshore transport causes EROSION or ACCRETION. With reference to winds or currents, the rate of increase or decrease in speed, usually in the vertical; or the curve that represents this rate. |
| Gravity | The attraction between bodies of matter is a gravity. |
| Current | Current refers to the flowing of water, or other liquid or gas. That portion of a stream of water which is moving with a velocity much greater than the average or in which the progress of the water is principally concentrated. Ocean currents can be classified in a number of different ways. |
| Aquatic | Aquatic refers to pertaining to water. |
| Dorsal | The upper or back surface of an animal is a dorsal. |
| Site | A factor considering the summation of all environmental features of a location that influences the placement of a city is a site. |
| Brine | Brine refers to water having a much higher. |
| Ventral | Ventral refers to of or pertaining to the underside of animals. |
| Reference point | Reference point refers to a specified location to which measurements are referred. In beach material studies, a specified point within the reference zone. |
| Point | Point refers to the extreme end of a cape, or the outer end of any land area protruding into the water, usually less prominent than a cape. A low profile shoreline promontory of more or less triangular shape, the top of which extends seaward. |
| Reach | Reach refers to an arm of the ocean extending into the land. A straight section of restricted waterway of considerable extent; may be similar to a narrows, except much longer in extent. |
| Population | Group of individual organisms of the same species living within a particular area is referred to as a population. |
| Observations | Information obtained through one or more of the five senses or through instruments that extend the senses are observations. |

# Chapter 10. Mechanisms of Orientation

## Chapter 10. Mechanisms of Orientation

| | |
|---|---|
| Population change | Population change refers to an increase or decrease in the size of a population. |
| Wind | Wind refers to the mass movement of air. |
| Desert | Desert refers to biome in which evaporation exceeds precipitation and the average amount of precipitation is less than 25 centimeters a year. Such areas have little vegetation or have widely spaced, mostly low vegetation. Compare forest, grassland. |
| Prey | Organism that is captured and serves as a source of food for an organism of another species is called prey. |
| Accuracy | Accuracy refers to the extent to which a measurement agrees with the accepted or correct value for that quantity, based on careful measurements by many people over a long time. Compare to precision. |
| Open sea | Same as high seas are referred to as the open sea. |
| Sea | Sea refers to the ocean. A large body of salt water, second in rank to an ocean, more or less landlocked and generally part of, or connected with, an ocean or a larger sea. State of the ocean or lake surface, in regard to waves. |
| Hierarchy | Hierarchy refers to grouping of objects by degrees of complexity, grade, or class. A hierarchical system of nomenclature is based on distinctions within groups and between groups. |
| System | A set of components that function and interact in some regular and theoretically predictable manner is called a system. |
| Landmark | Landmark refers to a conspicious object, natural or man-made, located near or on land, which aids in fixing the position of an observer. |
| Invertebrates | Invertebrates refers to animals that have no backbones. Compare vertebrates. |
| Lead | Lead refers to a heavy metal that is an important constituent of automobile batteries and other industrial products. A toxic metal capable of causing environmental disruption and producing a health problem to people and other living organisms. |
| Entrance | The entrance to a navigable BAY, HARBOR or CHANNEL, INLET or mouth separating the ocean from an inland water body. |
| Science | Attempts to discover order in nature and use that knowledge to make predictions about what should happen in nature is called science. |
| Horizon | Horizon refers to the line or circle which forms the apparent boundary between Earth and sky. A plane in rock strata characterized by particular features, as occurrence of distinctive fossil species. One of the series of distinctive layers found in a vertical cross-section of any well-developed soil. |
| Latitude | Latitude refers to distance from the equator. Compare altitude. |
| Altitude | Altitude refers to the height above sea level. Compare to latitude. |
| Fall | Fall refers to a mass moving nearly vertical and downward under the influence of gravity. |
| Glass | Glass refers to matter created when magma cools too quickly for atoms to arrange themselves into the ordered atomic structures of minerals. Most glasses are supercooled liquids. |
| Development | Development refers to change from a society that is largely rural, agricultural, illiterate, and poor, with a rapidly growing population, to one that is mostly urban, industrial, educated, and wealthy, with a slowly growing or stationary population. |
| Brines | With respect to mineral resources, refers to waters with a high salinity that contain useful |

# Chapter 10. Mechanisms of Orientation

## Chapter 10. Mechanisms of Orientation

| | |
|---|---|
| | materials such as bromine, iodine, calcium chloride, and magnesium, we have brines. |
| Well | Well refers to a hole, generally cylindrical and usually walled or lined with pipe, that is dug or drilled into the ground to penetrate an aquifer below the zone of saturation. |
| Wave | An oscillatory movement in a body of water manifested by an alternate rise and fall of the surface is called a wave. Disturbances of the surface of a liquid body, as the ocean, in the form of a ridge, swell or hump. The term wave by itself usually refers to the term surface gravity wave. |
| Atmosphere | Atmosphere refers to the whole mass of air surrounding the earth. |
| Degree | An arbitrary measure of temperature. One degree Celsius _ 1.8 degrees Fahrenheit. |
| Elevation | Elevation refers to the distance of a point above a specified surface of constant potential; the distance is measured along the direction of gravity between the point and the surface. |
| Substrate | A general term in reference to the surface on or within which organisms live is a substrate. |
| Disperse | Disperse refers to to spread or distribute from a fixed or constant source. To cause to become widely separated. |
| Shoreline | Shoreline refers to the intersection of a specified plane of water with the shore. All of the water areas of the state, including reservoirs and their associated uplands, together with the lands underlying them. |
| Predator | Predator refers to an organism that captures and feeds on parts or all of an organism of another species. |
| Shore | That strip of ground bordering any body of water which is alternately exposed, or covered by tides and/or waves is a shore. A shore of unconsolidated material is usually called a beach. |
| Compound | Compound refers to combination of atoms, or oppositely charged ions, of two or more different elements held together by attractive forces called chemical bonds. Compare element. |
| Output | Output refers to matter, energy, or information leaving a system. Compare with input, throughput. |
| Bar | An offshore ridge or mound of sand, gravel, or other unconsolidated material which is submerged, especially at the mouth of a river or estuary, or lying parallel to, and a short distance from, the beach is referred to as a bar. |
| Core | Core refers to a cylindrical sample extracted from a beach or seabed to investigate the types and DEPTHS of sediment layers. An inner, often much less permeable portion of a BREAKWATER, or BARRIER beach. |
| Magnetic pole | Magnetic pole refers to the point where the Earth's magnetic field flows back into the ground. Currently, this point is near the North Pole. |
| Force | Force refers to a push or pull that affects motion. The product of mass and acceleration of a material. |
| Dip | The angle of inclination measured in degrees from the horizontal is called dip. |
| Direction of current | Direction toward which CURRENT is flowing is called direction of current. |
| Energy | Capacity to do work by performing mechanical, physical, chemical, or electrical tasks or to cause a heat transfer between two objects at different temperatures is an energy. |
| Range | Land used for grazing is referred to as the range. |
| Ocean | The great body of salt water which occupies two-thirds of the surface of the Earth, or one of |

Go to **Cram101.com** for the Practice Tests for this Chapter.

## Chapter 10. Mechanisms of Orientation

| | |
|---|---|
| | its major subdivisions is called an ocean. |
| Coast | A strip of land of indefinite length and width that extends from the SEASHORE inland to the first major change in terrain features is referred to as coast. |
| Sargasso Sea | Sargasso sea refers to the central portion of the North Atlantic circulation gyre, which contains large quantities of the floating brown alga Sargassum. |
| Offshore | Offshore in beach terminology refers to the comparatively flat zone of variable width, extending from the shoreface to the edge of the continental shelf. It is continually submerged. The direction seaward from the shore. The zone beyond the nearshore zone where sediment motion induced by waves alone effectively ceases and where the influence of the sea bed on wave action is small in comparison with the effect of wind. The breaker zone directly seaward of the low tide line. |
| Recent | A synonym of Holocene is called recent. |
| Interference | Interference refers to during succession, one species prevents the entrance of later successional species into an ecosystem. For example, some grasses produce such dense and thick mats that seeds of trees cannot reach the soil to germinate. As long as these grasses persist, th. |
| Topography | Topography refers to the form of the features of the actual surface of the Earth in a particular region considered collectively. |
| Storm | Local or regional atmospheric disturbance characterized by strong winds often accompanied by precipitation is referred to as a storm. |
| Magnetism | Magnetism refers to a group of physical phenomena associated with moving electricity. |
| Soil | Soil refers to a layer of weathered, unconsolidated material on top of bedrock; often also defined as containing organic matter and being capable of supporting plant growth. |
| Plants | Eukaryotic, mostly multicelled organisms such as algae, mosses, ferns, flowers, cacti, grasses, beans, wheat, rice, and trees are plants. These organisms use photosynthesis to produce organic nutrients for themselves and for other organisms is referred to as plants. |
| Hypothesis | In science, an explanation set forth in a manner that can be tested and is capable of being disproved. A tested hypothesis is accepted until and unless it has been disproved. |
| Effluent | Effluent refers to any material that flows outward from something. Examples include wastewater from hydroelectric plants and water discharged into streams from waste-disposal sites. |
| Reduce | With respect to waste management, reduce refers to practices that will reduce the amount of waste we produce. |
| Forest | Forest refers to biome with enough average annual precipitation to support growth of various species of trees and smaller forms of vegetation. Compare desert, grassland. |
| Tests | The skeleton or shells of certain microorganisms are called tests. |
| Valley | An elongated depression, usually with an outlet, between bluffs or between ranges of hills or mountains is a valley. |
| Thunderstorm | Thunderstorm refers to a tall, buoyant cloud of moist air that generates lightning and thunder usually accompanied by rain, gusty winds, and sometimes hail. |
| Frequency | Number of events in a given time interval. For earthquakes, it is the number of cycles of seismic waves that pass in a second; frequency = l/period. |
| Situation | The relative geographic location of a site that makes it a good location for a city is a |

Go to **Cram101.com** for the Practice Tests for this Chapter.

## Chapter 10. Mechanisms of Orientation

| | |
|---|---|
| | situation. |
| **Disturbance** | Disturbance refers to a discrete event in time that disrupts an ecosystem or community. Examples of natural disturbances include fires, hurricanes, tornadoes, droughts, and floods. Examples of humancaused disturbances include deforestation, overgrazing, and plowing. |
| **Tides** | The periodic rise and fall of the Earth's water surface as a consequence of the gravitational attraction of the Moon and the Sun, which are called tides. |
| **Organism** | Any form of life is referred to as an organism. |
| **Rock** | Any material that makes up a large, natural, continuous part of earth's crust is rock. |
| **Amplitude** | Half of the peak-to-trough range of a wave is the amplitude. |
| **Input** | Input refers to matter, energy, or information entering a system. Compare output, throughput. |
| **Trough** | A long and broad submarine depression with gently sloping sides is referred to as a trough. |
| **Ecology** | Ecology refers to study of the interactions of living organisms with one another and with their nonliving environment of matter and energy; study of the structure and functions of nature. |

# Chapter 10. Mechanisms of Orientation

## Chapter 11. The Ecology and Evolution of Spatial Distribution

| | |
|---|---|
| Benefits | Benefits refers to the economic value of a scheme, usually measured in terms of the cost of damages avoided by the scheme, or the valuation of perceived amenity or environmental improvements. |
| Ecology | Ecology refers to study of the interactions of living organisms with one another and with their nonliving environment of matter and energy; study of the structure and functions of nature. |
| Migration | A term that refers to the habit of some animals is a migration. |
| Spring | A place where groundwater flows out onto the surface is a spring. |
| Head | A comparatively high promontory with either a CLIFF or steep face. It extends into a large body of water, such as a sea or lake. An unnamed HEAD is usually called a headland. The section of RIP CURRENT which has widened out seaward of the BREAKERS, also called head of |
| Reach | Reach refers to an arm of the ocean extending into the land. A straight section of restricted waterway of considerable extent; may be similar to a narrows, except much longer in extent. |
| Range | Land used for grazing is referred to as the range. |
| Site | A factor considering the summation of all environmental features of a location that influences the placement of a city is a site. |
| Disperse | Disperse refers to to spread or distribute from a fixed or constant source. To cause to become widely separated. |
| Conditions | Conditions refers to physical or chemical attributes of the environment that, while not being consumed, influence biological processes and population growth. Examples are temperature, salinity, and acidity. Compare resources. |
| Point | Point refers to the extreme end of a cape, or the outer end of any land area protruding into the water, usually less prominent than a cape. A low profile shoreline promontory of more or less triangular shape, the top of which extends seaward. |
| Species | Group of organisms that resemble one another in appearance, behavior, chemical makeup and processes, and genetic structure is a species. Organisms that reproduce sexually are classified as members of the same species only if they can breed with one another and produce offspring. |
| Lead | Lead refers to a heavy metal that is an important constituent of automobile batteries and other industrial products. A toxic metal capable of causing environmental disruption and producing a health problem to people and other living organisms. |
| Risk | Risk refers to the probability that something undesirable will happen from deliberate or accidental exposure. |
| Alleles | Slightly different molecular forms found in a particular gene are alleles. |
| Population | Group of individual organisms of the same species living within a particular area is referred to as a population. |
| Frequency | Number of events in a given time interval. For earthquakes, it is the number of cycles of seismic waves that pass in a second; frequency = l/period. |
| Competition | Two or more individual organisms of a single species or two or more individuals of different species attempting to use the same scarce resources in the same ecosystem is a competition. |
| Food | General term for organic molecules capable of providing energy to heterotrophs when combined with oxygen during biochemical respiration is called food. |
| Resources | Resources refer to substances that can be consumed by an organism and, as a result, become |

Go to **Cram101.com** for the Practice Tests for this Chapter.

# Chapter 11. The Ecology and Evolution of Spatial Distribution

## Chapter 11. The Ecology and Evolution of Spatial Distribution

| | |
|---|---|
| | unavailable to other organisms. |
| Genes | Genes refers to coded units of information about specific traits that are passed on from parents to offspring during reproduction. They consist of segments of DNA molecules found in chromosomes. |
| Well | Well refers to a hole, generally cylindrical and usually walled or lined with pipe, that is dug or drilled into the ground to penetrate an aquifer below the zone of saturation. |
| Reduce | With respect to waste management, reduce refers to practices that will reduce the amount of waste we produce. |
| Stress | Stress refers to force per unit area. May be compression, tension, or shear. |
| Energy | Capacity to do work by performing mechanical, physical, chemical, or electrical tasks or to cause a heat transfer between two objects at different temperatures is an energy. |
| Predation | Predation refers to a situation in which an organism of one species captures and feeds on parts or all of an organism of another species. |
| Probability | A mathematical statement about how likely it is that something will happen is probability. |
| Hypothesis | In science, an explanation set forth in a manner that can be tested and is capable of being disproved. A tested hypothesis is accepted until and unless it has been disproved. |
| System | A set of components that function and interact in some regular and theoretically predictable manner is called a system. |
| Force | Force refers to a push or pull that affects motion. The product of mass and acceleration of a material. |
| Dispersion | Dispersion refers to act of dispersing, or state of being dispersed. The separation of waves by virtue of their differing rates of travel. |
| Reproduction | Production of offspring by one is called reproduction. |
| Parasitism | Interaction between species in which one organism, called the parasite, preys on another organism, called the host, by living on or in the host is referred to as parasitism. |
| Resource | Resource refers to anything obtained from the living and nonliving environment to meet human needs and wants. It can also be applied to other species. |
| Degree | An arbitrary measure of temperature. One degree Celsius _ 1.8 degrees Fahrenheit. |
| Plants | Eukaryotic, mostly multicelled organisms such as algae, mosses, ferns, flowers, cacti, grasses, beans, wheat, rice, and trees are plants. These organisms use photosynthesis to produce organic nutrients for themselves and for other organisms is referred to as plants. |
| Bank | The rising ground bordering a lake, river or sea is referred to as a bank. |
| Experiment | Experiment refers to procedure a scientist uses to study some phenomenon under known conditions. Some experiments are conducted in the laboratory, but others are conducted in nature. The resulting scientific data or facts must be verified or confirmed by repeated observation. |
| Fall | Fall refers to a mass moving nearly vertical and downward under the influence of gravity. |
| Clumped distribution | Clumped distribution refers to distribution of organisms within a community in small, patchy aggregations, or clumps; the most common distribution pattern. |
| Mass | The amount of material in an object is the mass. |
| Coast | A strip of land of indefinite length and width that extends from the SEASHORE inland to the first major change in terrain features is referred to as coast. |

# Chapter 11. The Ecology and Evolution of Spatial Distribution

## Chapter 11. The Ecology and Evolution of Spatial Distribution

| | |
|---|---|
| Tundra | The treeless land area in alpine and arctic areas characterized by plants of low stature and including bare areas without any plants and covered areas with lichens, mosses, grasses, sedges, and small flowering plants, including low shrubs are referred to as the tundra. |
| Fat | Fat refers to an organic compound composed of carbon, hydrogen, and oxygen that is insoluble in water. |
| Sea | Sea refers to the ocean. A large body of salt water, second in rank to an ocean, more or less landlocked and generally part of, or connected with, an ocean or a larger sea. State of the ocean or lake surface, in regard to waves. |
| Natural selection | Process by which a particular beneficial gene is reproduced more than other genes in succeeding generations is natural selection. The result of natural selection is a population that contains a greater proportion of organisms better adapted to certain environments. |
| Protein | Protein refers to a family of complex organic compounds containing nitrogen and composed of various amino acids. |
| Reserves | Reserves refer to resources that have been identified from which a usable mineral can be extracted profitably at present prices with current mining. |
| Fuel | Air,/ substar ce that produces heat by combustion is a fuel. |
| Predator | Predator refers to an organism that captures and feeds on parts or all of an organism of another species. |
| Prey | Organism that is captured and serves as a source of food for an organism of another species is called prey. |
| Terrestrial | Terrestrial refers to pertaining to land. Compare with aquatic. |
| Matter | Matter refers to anything that has mass and takes up space. On earth, where gravity is present, we weigh an object to determine its mass. |
| Weather | Weather refers to short-term changes in the temperature, barometric pressure, humidity, precipitation, sunshine, cloud cover, wind direction and speed, and other conditions in the troposphere at a given place and time. Compare with climate. |
| Lake | Large natural body of standing fresh water formed when water from precipitation, land runoff, or groundwater flow fills a depression in the earth created by glaciation, earth movement, volcanic activity, or a giant meteorit are called the lake. |
| Storm | Local or regional atmospheric disturbance characterized by strong winds often accompanied by precipitation is referred to as a storm. |
| Tornado | Tornado refers to a localized, narrow, violent funnel of fast-spinning wind, usually generated when two air masses collide. Not to be confused with a cyclone. |
| Gulf | A relatively large portion of sea, partly enclosed by land is called gulf. |
| Valley | An elongated depression, usually with an outlet, between bluffs or between ranges of hills or mountains is a valley. |
| Habitat | The place where an organism lives is called habitat. |
| Temperature | Temperature refers to a measure of the average speed of motion of the atoms, ions, or molecules in a substance or combination of substances at a given moment. Compare with heat. |
| Forest | Forest refers to biome with enough average annual precipitation to support growth of various species of trees and smaller forms of vegetation. Compare desert, grassland. |
| Thinning | The timber harvesting practice of selectively removing only smaller or poorly formed trees is called thinning. |

Go to **Cram101.com** for the Practice Tests for this Chapter.

## Chapter 11. The Ecology and Evolution of Spatial Distribution

## Chapter 11. The Ecology and Evolution of Spatial Distribution

| | |
|---|---|
| **Dip** | The angle of inclination measured in degrees from the horizontal is called dip. |
| **Climate** | Physical properties of the troposphere of an area based on analysis of its weather records over a long period. The two main factors determining an area's climate are temperature, with its seasonal variations, and the amount and distri. |
| **Adaptation** | Any genetically controlled structural, physiological, or behavioral characteristic that helps an organism survive and reproduce under a given set of environmental conditions. It usually results from a beneficial mutation is an adaptation. |
| **Response** | The amount of health damage caused by exposure to a certain dose of a harmful substance or form of radiation is a response. |
| **Zone** | Division or province of the ocean with homogeneous characteristics is referred to as a zone. |
| **Annual** | Annual refers to plant that grows, sets seed, and dies in one growing season. Compare to perennial. |
| **Microorganisms** | Organisms that are so small that they can be seen only by using a microscope are referred to as microorganisms. |
| **Fertility** | Fertility refers to the ability to produce offspring; the proportion of births to population. |
| **Gene** | Gene refers to the fundamental unit in inheritance; it carries the characteristics of parents to their offspring. |
| **Environment** | Environment refers to all external conditions and factors, living and nonliving, that affect an organism or other specified system during its lifetime; the earth's life-support systems for us and for all other forms of life-another term for solar capita. |
| **Net energy** | Net energy refers to the total amount of useful energy available from an energy resource or energy system over its lifetime minus the amount of energy used, automatically wasted, and unnecessarily wasted in finding, and processing. |

## Chapter 12. Foraging

| | |
|---|---|
| Food | General term for organic molecules capable of providing energy to heterotrophs when combined with oxygen during biochemical respiration is called food. |
| Predation | Predation refers to a situation in which an organism of one species captures and feeds on parts or all of an organism of another species. |
| Competition | Two or more individual organisms of a single species or two or more individuals of different species attempting to use the same scarce resources in the same ecosystem is a competition. |
| Rock | Any material that makes up a large, natural, continuous part of earth's crust is rock. |
| Prey | Organism that is captured and serves as a source of food for an organism of another species is called prey. |
| Plants | Eukaryotic, mostly multicelled organisms such as algae, mosses, ferns, flowers, cacti, grasses, beans, wheat, rice, and trees are plants. These organisms use photosynthesis to produce organic nutrients for themselves and for other organisms is referred to as plants. |
| Food web | Complex network of many interconnected food chains and feeding relationships. Compare food chai is called food web. |
| Community | Community refers to populations of all species living and interacting in an area at a particular time. |
| Pile | A long substantial pole of wood, concrete or metal, driven into the earth or sea bed to serve as a support or protection is called pile. |
| Density | Density refers to the ratio of a mass to a unit volume specified as grams per cubic centimeter. |
| Biomass | Biomass refers to organic matter produced by plants and other photosynthetic producers; total dry weight of all living organisms that can be supported at each trophic level in a food chain or web; dry weight of all organic matter in plants and animals in an ecosystem; pla. |
| Well | Well refers to a hole, generally cylindrical and usually walled or lined with pipe, that is dug or drilled into the ground to penetrate an aquifer below the zone of saturation. |
| Energy | Capacity to do work by performing mechanical, physical, chemical, or electrical tasks or to cause a heat transfer between two objects at different temperatures is an energy. |
| River | A natural stream of water larger than a brook or creek is a river. |
| Tundra | The treeless land area in alpine and arctic areas characterized by plants of low stature and including bare areas without any plants and covered areas with lichens, mosses, grasses, sedges, and small flowering plants, including low shrubs are referred to as the tundra. |
| Forest | Forest refers to biome with enough average annual precipitation to support growth of various species of trees and smaller forms of vegetation. Compare desert, grassland. |
| Desert | Desert refers to biome in which evaporation exceeds precipitation and the average amount of precipitation is less than 25 centimeters a year. Such areas have little vegetation or have widely spaced, mostly low vegetation. Compare forest, grassland. |
| Ocean | The great body of salt water which occupies two-thirds of the surface of the Earth, or one of its major subdivisions is called an ocean. |
| Organism | Any form of life is referred to as an organism. |
| Species | Group of organisms that resemble one another in appearance, behavior, chemical makeup and processes, and genetic structure is a species. Organisms that reproduce sexually are classified as members of the same species only if they can breed with one another and produce offspring. |

# Chapter 12. Foraging

## Chapter 12. Foraging

| | |
|---|---|
| Aquatic | Aquatic refers to pertaining to water. |
| Mollusks | The category of animals that includes chitons, snails, clams, and octopuses is referred to as mollusks. |
| Environment | Environment refers to all external conditions and factors, living and nonliving, that affect an organism or other specified system during its lifetime; the earth's life-support systems for us and for all other forms of life-another term for solar capita. |
| Invertebrates | Invertebrates refers to animals that have no backbones. Compare vertebrates. |
| Vertebrates | Vertebrates refer to animals with backbones. Compare with invertebrates. |
| Vertebrate | A chordate with a segmented backbone is a vertebrate. |
| Mud | Mud refers to a mixture of silt and clay sized particles. |
| Baleen | Baleen refers to a series of elastic, horny plates that grow in place of teeth in the upper jaw of a certain group of whales called Mysticeti. |
| Plates | Various-sized areas of earth's lithosphere that move slowly around with the mantle's flowing asthenosphere are plates. Most earthquakes and volcanoes occur around the boundaries of these plates. |
| Strain | A change in torn or size of a body due to external forces is called strain. |
| Zooplankton | Zooplankton refers to animal plankton. Small floating herbivores that feed on plant plankton. Compare with phytoplankton. |
| Population | Group of individual organisms of the same species living within a particular area is referred to as a population. |
| Microorganisms | Organisms that are so small that they can be seen only by using a microscope are referred to as microorganisms. |
| Nourishment | Nourishment refers to the process of replenishing a beach. It may be brought about naturally, by longshore transport, or artificially by the deposition of dredged materials. |
| Surge | Surge refers to long-interval variations in velocity and pressure in fluid flow, not necessarily periodic, perhaps even transient in nature. The name applied to wave motion with a period intermediate between that of an ordinary wind wave and that of the tide. Changes in water level as a result of meteorological forcing causing a difference between the recorded water level and that predicted using harmonic analysis, may be positive or negative. |
| Forcing | With respect to global change, processes capable of changing global temperature, such as changes in solar energy emitted from the sun, or volcanic activity, we have forcing. |
| Mist | Water vapor suspended in the air in very small drops finer than rain, larger than fog is referred to as mist. |
| Krill | Euphausia superba, a thumb-sized crustacean common in Antarctic waters is called krill. |
| Frequency | Number of events in a given time interval. For earthquakes, it is the number of cycles of seismic waves that pass in a second; frequency = l/period. |
| Habitat | The place where an organism lives is called habitat. |
| Conditions | Conditions refers to physical or chemical attributes of the environment that, while not being consumed, influence biological processes and population growth. Examples are temperature, salinity, and acidity. Compare resources. |
| Ultraviolet | The longest wavelength of ultraviolet radiation, not affected by stratospheric ozone, and is transmitted to the surface of the earth is ultraviolet. |

## Chapter 12. Foraging

## Chapter 12. Foraging

| | |
|---|---|
| Degree | An arbitrary measure of temperature. One degree Celsius _ 1.8 degrees Fahrenheit. |
| Predator | Predator refers to an organism that captures and feeds on parts or all of an organism of another species. |
| Tongue | A long narrow strip of land, projecting into a body of water is a tongue. |
| Head | A comparatively high promontory with either a CLIFF or steep face. It extends into a large body of water, such as a sea or lake. An unnamed HEAD is usually called a headland. The section of RIP CURRENT which has widened out seaward of the BREAKERS, also called head of |
| Reach | Reach refers to an arm of the ocean extending into the land. A straight section of restricted waterway of considerable extent; may be similar to a narrows, except much longer in extent. |
| Fungi | Fungi refers to eukaryotic, mostly multicelled organisms such as mushrooms, molds, and yeasts. As decomposers, they get the nutrients they need by secreting enzymes that speed up the breakdown the organic matter in the tissue of other living or dead organisms. Then they. |
| Bacteria | Bacteria refer to prokaryotic, one-celled organisms. Some transmit diseases. Most act as decomposers and get the nutrients they need by breaking down complex organic compounds in the tissues of living or dead organisms into simpler inorganic nutrient compounds. |
| Coral reef | Formation produced by massive colonies containing billions of tiny coral animals, called polyps, that secrete a stony substance around themselves for protection. When the corals die, their empty outer skeletons form layers that caus is referred to as coral reef. |
| Reef | A ridge of rock or other material lying just below the surface of the sea is called a reef. |
| Rocks | An aggregate of one or more minerals rather large in area are rocks. The three classes of rocks are the following: Igneous rock - crystalline rocks formed from molten material. Sedimentary rock - A rock resulting from the consolidation of loose sediment that has accumulated in layers. Metamorphic rock - Rock that has formed from preexisting rock as a result of heat or pressure. |
| Tests | The skeleton or shells of certain microorganisms are called tests. |
| Disturbance | Disturbance refers to a discrete event in time that disrupts an ecosystem or community. Examples of natural disturbances include fires, hurricanes, tornadoes, droughts, and floods. Examples of humancaused disturbances include deforestation, overgrazing, and plowing. |
| Wind | Wind refers to the mass movement of air. |
| Kingdom | Kingdom refers to the largest category of biological classification. Five kingdoms are presently recognized. |
| Heat | Total kinetic energy of all the randomly moving atoms, ions, or molecules within a given substance, excluding the overall motion of the whole object. This form of kinetic energy flows from one body to another when there is a temperature difference betwe is referred to as heat. |
| Infrared radiation | The part of the electromagnetic spectrum that we perceive as heat is referred to as infrared radiation. |
| Radiation | Fast-moving particles or waves of energy are called radiation. |
| Temperature | Temperature refers to a measure of the average speed of motion of the atoms, ions, or molecules in a substance or combination of substances at a given moment. Compare with heat. |
| Range | Land used for grazing is referred to as the range. |
| Accuracy | Accuracy refers to the extent to which a measurement agrees with the accepted or correct value for that quantity, based on careful measurements by many people over a long time. |

## Chapter 12. Foraging

| | |
|---|---|
| | Compare to precision. |
| Sand | Sand refers to an unconsolidated mixture of inorganic soil consisting of small but easily distinguishable grains ranging in size from about .062 mm to 2.0 mm. |
| Mole | In coastal terminology, a massive solid-filled structure of earth, masonry or large stone is referred to as a mole. |
| System | A set of components that function and interact in some regular and theoretically predictable manner is called a system. |
| Wetlands | Lands whose saturation with water is the dominant factor determining the nature of soil development and the types of plant and animal communities that live in the soil and on its surface are wetlands. |
| Base | A substance that combines with a hydrogen ion in solution is called the base. |
| Hot spot | A zone on a lithospheric plate that overlies unusually hot asthenosphere, and where large volumes of lava commonly are extruded, building a large volcanic pile on the sea floor is called hot spot. |
| Volatile | Substances that readily become gases when pressure is decreased, or temperature increased are referred to as volatile. |
| Phytoplankton | Phytoplankton refers to small, drifting plants, mostly algae and bacteria, found in aquatic ecosystems. Compare with plankton, zooplankton. |
| Lead | Lead refers to a heavy metal that is an important constituent of automobile batteries and other industrial products. A toxic metal capable of causing environmental disruption and producing a health problem to people and other living organisms. |
| Cape | Cape refers to a relatively extensive land area jutting seaward from a continent or large island which prominently marks a change in, or interrupts notably, the coastal trend; a prominent feature. |
| Storm | Local or regional atmospheric disturbance characterized by strong winds often accompanied by precipitation is referred to as a storm. |
| Substrate | A general term in reference to the surface on or within which organisms live is a substrate. |
| Chemical | One of the millions of different elements and compounds found naturally or synthesized by human is referred to as chemical. |
| Dorsal | The upper or back surface of an animal is a dorsal. |
| Probability | A mathematical statement about how likely it is that something will happen is probability. |
| Slide | In mass wasting, movement of a descending mass along a plane approximately parallel to the slope of the surface is referred to as slide. |
| Key | Key refers to a low, insular BANK of sand, coral, etc., as one of the islets off the southern coast of Florida. |
| Experiment | Experiment refers to procedure a scientist uses to study some phenomenon under known conditions. Some experiments are conducted in the laboratory, but others are conducted in nature. The resulting scientific data or facts must be verified or confirmed by repeated observation. |
| Hypothesis | In science, an explanation set forth in a manner that can be tested and is capable of being disproved. A tested hypothesis is accepted until and unless it has been disproved. |
| Response | The amount of health damage caused by exposure to a certain dose of a harmful substance or form of radiation is a response. |

# Chapter 12. Foraging

## Chapter 12. Foraging

| | |
|---|---|
| Recent | A synonym of Holocene is called recent. |
| Efficiency | The ratio of output to input. With machines, usually the ratio of work or power produced to the energy or power used to operate or fuel them. With living things, efficiency may be defined as either the useful work done or the energy stored in a useful fo. |
| Situation | The relative geographic location of a site that makes it a good location for a city is a situation. |
| Theory | A general explanation of a characteristic of nature consistently supported by observation or experiment is referred to as a theory. |
| Natural selection | Process by which a particular beneficial gene is reproduced more than other genes in succeeding generations is natural selection. The result of natural selection is a population that contains a greater proportion of organisms better adapted to certain environments. |
| Benefits | Benefits refers to the economic value of a scheme, usually measured in terms of the cost of damages avoided by the scheme, or the valuation of perceived amenity or environmental improvements. |
| Net energy | Net energy refers to the total amount of useful energy available from an energy resource or energy system over its lifetime minus the amount of energy used, automatically wasted, and unnecessarily wasted in finding, and processing. |
| Matter | Matter refers to anything that has mass and takes up space. On earth, where gravity is present, we weigh an object to determine its mass. |
| Reproduction | Production of offspring by one is called reproduction. |
| Seacoast | Seacoast refers to the coast adjacent to the sea or ocean. |
| Tide | The periodic rising and falling of the water that results from gravitational attraction of the moon and sun acting upon the rotating earth is the tide. Although the accompanying horizontal movement of the water resulting from the same cause is also sometimes called the tide, it is preferable to designate the latter as tidal current, reserving the name tide for the vertical movement. |
| Shore | That strip of ground bordering any body of water which is alternately exposed, or covered by tides and/or waves is a shore. A shore of unconsolidated material is usually called a beach. |
| Sea | Sea refers to the ocean. A large body of salt water, second in rank to an ocean, more or less landlocked and generally part of, or connected with, an ocean or a larger sea. State of the ocean or lake surface, in regard to waves. |
| Wave | An oscillatory movement in a body of water manifested by an alternate rise and fall of the surface is called a wave. Disturbances of the surface of a liquid body, as the ocean, in the form of a ridge, swell or hump. The term wave by itself usually refers to the term surface gravity wave. |
| Observations | Information obtained through one or more of the five senses or through instruments that extend the senses are observations. |
| Risk | Risk refers to the probability that something undesirable will happen from deliberate or accidental exposure. |
| Threshold | Threshold refers to a point in the operation of a system at which a change occurs. With respect to toxicology, it is a level below which effects are not observable and above which effects become apparent. |
| Nutrient | Any atom, ion, or molecule an organism that needs to live, grow, or reproduce is a nutrient. |
| Mixture | Mixture refers to combination of two or more elements and compounds. |

# Chapter 12. Foraging

## Chapter 12. Foraging

| | |
|---|---|
| Lake | Large natural body of standing fresh water formed when water from precipitation, land runoff, or groundwater flow fills a depression in the earth created by glaciation, earth movement, volcanic activity, or a giant meteorit are called the lake. |
| Concentration | Amount of a chemical in a particular volume or weight of air, water, soil, or other medium is referred to as concentration. |
| Site | A factor considering the summation of all environmental features of a location that influences the placement of a city is a site. |
| Site quality | Used by foresters to mean an estimator of the maximum timber crop the land can produce in a given time is referred to as site quality. |
| Bar | An offshore ridge or mound of sand, gravel, or other unconsolidated material which is submerged, especially at the mouth of a river or estuary, or lying parallel to, and a short distance from, the beach is referred to as a bar. |
| Reduce | With respect to waste management, reduce refers to practices that will reduce the amount of waste we produce. |
| Current | Current refers to the flowing of water, or other liquid or gas. That portion of a stream of water which is moving with a velocity much greater than the average or in which the progress of the water is principally concentrated. Ocean currents can be classified in a number of different ways. |
| Reserves | Reserves refer to resources that have been identified from which a usable mineral can be extracted profitably at present prices with current mining. |
| Point | Point refers to the extreme end of a cape, or the outer end of any land area protruding into the water, usually less prominent than a cape. A low profile shoreline promontory of more or less triangular shape, the top of which extends seaward. |
| Resources | Resources refer to substances that can be consumed by an organism and, as a result, become unavailable to other organisms. |
| Travel time | The time necessary for waves to travel a given distance from the generating area is travel time. |
| Tension | A state of stress that tends to pull the body apart is called tension. |
| Gravel | Gravel refers to loose, rounded fragments of rock, larger than sand, but smaller than cobbles. Small stones and pebbles, or a mixture of these with sand. |
| Water table | The upper surface of a zone of saturation, where the body of groundwater is not confined by an overlying impermeable formation is a water table. Where an overlying confining formation exists, the aquifer in question has no water table. |
| Bed | The bottom of a watercourse, or any body of water is called a bed. |
| Parasitism | Interaction between species in which one organism, called the parasite, preys on another organism, called the host, by living on or in the host is referred to as parasitism. |
| Core | Core refers to a cylindrical sample extracted from a beach or seabed to investigate the types and DEPTHS of sediment layers. An inner, often much less permeable portion of a BREAKWATER, or BARRIER beach. |

## Chapter 12. Foraging

## Chapter 13. Antipredator Behavior

| | |
|---|---|
| Chemical | One of the millions of different elements and compounds found naturally or synthesized by human is referred to as chemical. |
| Prey | Organism that is captured and serves as a source of food for an organism of another species is called prey. |
| Predation | Predation refers to a situation in which an organism of one species captures and feeds on parts or all of an organism of another species. |
| Food | General term for organic molecules capable of providing energy to heterotrophs when combined with oxygen during biochemical respiration is called food. |
| Plants | Eukaryotic, mostly multicelled organisms such as algae, mosses, ferns, flowers, cacti, grasses, beans, wheat, rice, and trees are plants. These organisms use photosynthesis to produce organic nutrients for themselves and for other organisms is referred to as plants. |
| Poison | A chemical that in one dose kills exactly 50% of the animals in a test population within a 14-day period is referred to as poison. |
| Species | Group of organisms that resemble one another in appearance, behavior, chemical makeup and processes, and genetic structure is a species. Organisms that reproduce sexually are classified as members of the same species only if they can breed with one another and produce offspring. |
| System | A set of components that function and interact in some regular and theoretically predictable manner is called a system. |
| Well | Well refers to a hole, generally cylindrical and usually walled or lined with pipe, that is dug or drilled into the ground to penetrate an aquifer below the zone of saturation. |
| Predator | Predator refers to an organism that captures and feeds on parts or all of an organism of another species. |
| Fall | Fall refers to a mass moving nearly vertical and downward under the influence of gravity. |
| Risk | Risk refers to the probability that something undesirable will happen from deliberate or accidental exposure. |
| Site | A factor considering the summation of all environmental features of a location that influences the placement of a city is a site. |
| Environment | Environment refers to all external conditions and factors, living and nonliving, that affect an organism or other specified system during its lifetime; the earth's life-support systems for us and for all other forms of life-another term for solar capita. |
| Range | Land used for grazing is referred to as the range. |
| Marsh | Marsh refers to a tract of soft, wet land, usually vegetated by reeds, grasses and occasionally small shrubs. Soft, wet area periodically or continuously flooded to a shallow depth, usually characterized by a particular subclass of grasses, cattails and other low plants. |
| Forest | Forest refers to biome with enough average annual precipitation to support growth of various species of trees and smaller forms of vegetation. Compare desert, grassland. |
| Ventral | Ventral refers to of or pertaining to the underside of animals. |
| Dorsal | The upper or back surface of an animal is a dorsal. |
| Adaptation | Any genetically controlled structural, physiological, or behavioral characteristic that helps an organism survive and reproduce under a given set of environmental conditions. It usually results from a beneficial mutation is an adaptation. |

# Chapter 13. Antipredator Behavior

## Chapter 13. Antipredator Behavior

| | |
|---|---|
| Ultraviolet | The longest wavelength of ultraviolet radiation, not affected by stratospheric ozone, and is transmitted to the surface of the earth is ultraviolet. |
| Radiation | Fast-moving particles or waves of energy are called radiation. |
| Ocean | The great body of salt water which occupies two-thirds of the surface of the Earth, or one of its major subdivisions is called an ocean. |
| High water | High water refers to maximum height reached by a rising tide. The height may be solely due to the periodic tidal forces or it may have superimposed upon it the effects of prevailing meteorological conditions. Nontechnically, also called the HIGH TIDE. |
| Aquatic | Aquatic refers to pertaining to water. |
| Terrestrial | Terrestrial refers to pertaining to land. Compare with aquatic. |
| Cephalopod | The group of marine predators that includes squid, octopuses, and nautiluses is referred to as cephalopod. |
| Substrate | A general term in reference to the surface on or within which organisms live is a substrate. |
| Matter | Matter refers to anything that has mass and takes up space. On earth, where gravity is present, we weigh an object to determine its mass. |
| Wavelength | The horizontal distance between corresponding points on successive waves, such as from crest to crest or trough to trough is called a wavelength. |
| Larva | The immature form of an animal that differs significantly from the adult form is a larva. |
| Savanna | Savanna refers to an area with trees scattered widely among dense grasses. |
| Fire | The rapid combination of oxygen with organic material to produce flame, heat, and light is referred to as the fire. |
| Habitat | The place where an organism lives is called habitat. |
| Sand | Sand refers to an unconsolidated mixture of inorganic soil consisting of small but easily distinguishable grains ranging in size from about .062 mm to 2.0 mm. |
| Situation | The relative geographic location of a site that makes it a good location for a city is a situation. |
| Head | A comparatively high promontory with either a CLIFF or steep face. It extends into a large body of water, such as a sea or lake. An unnamed HEAD is usually called a headland. The section of RIP CURRENT which has widened out seaward of the BREAKERS, also called head of |
| Frequency | Number of events in a given time interval. For earthquakes, it is the number of cycles of seismic waves that pass in a second; frequency = l/period. |
| State | State refers to an expression of the internal form of matter. Water exists in three states: solid, liquid, and gas. A solid has a fixed volume and fixed shape; a liquid has a fixed volume but no fixed shape; and a gas has neither fixed volume nor fixed shape. |
| Gulf | A relatively large portion of sea, partly enclosed by land is called gulf. |
| Point | Point refers to the extreme end of a cape, or the outer end of any land area protruding into the water, usually less prominent than a cape. A low profile shoreline promontory of more or less triangular shape, the top of which extends seaward. |
| Degree | An arbitrary measure of temperature. One degree Celsius _ 1.8 degrees Fahrenheit. |
| Pollution | Pollution refers to an undesirable change in the physical, chemical, or biological characteristics of air, water, soil, or food that can adversely affect the health, survival, or activities of humans or other living organisms. |

# Chapter 13. Antipredator Behavior

## Chapter 13. Antipredator Behavior

| | |
|---|---|
| Population | Group of individual organisms of the same species living within a particular area is referred to as a population. |
| Air pollution | One or more chemicals in high enough concentrations in the air to harm humans, other animals, vegetation, or materials are called air pollution. Excess heat and noise can also be considered forms of air pollution. |
| Sulfur dioxide | Sulfur dioxide refers to a colorless and odorless gas normally present at the Earth's surface in low concentrations. An important precursor to acid rain. Major anthropogenic source is burning fossil fuels. |
| Economy | Economy refers to system of production, distribution, and consumption of economic goods. |
| Coal | Solid, combustible mixture of organic compounds with 30-98% carbon by weight, mixed with various amounts of water and small amounts of sulfur and nitrogen compounds. It is formed in several stages as the remains of plants are subjected to heat and press is a coal. |
| Lead | Lead refers to a heavy metal that is an important constituent of automobile batteries and other industrial products. A toxic metal capable of causing environmental disruption and producing a health problem to people and other living organisms. |
| Industrial melanism | A form of natural selection where the color of a living thing helps it to blend in with an urban, industrial environment is an industrial melanism. |
| Observations | Information obtained through one or more of the five senses or through instruments that extend the senses are observations. |
| Tolerance | The ability to withstand stress resulting from exposure to a pollutant or harmful condition is tolerance. |
| Pollutant | A particular chemical or form of energy that can adversely affect the health, survival, or activities of humans or other living organism is a pollutant. |
| Recent | A synonym of Holocene is called recent. |
| Granite | A light-colored, coarsegrained, intrusive igneous rock composed mainly of quartz and feldspar and that typifies the continental crust are called the granite. |
| Algae | Algae refers to simple marine and freshwater plants, unicellular and multicellular, that lack roots, stems, and leaves. |
| Gravity | The attraction between bodies of matter is a gravity. |
| Stream | Stream refers to any flow of water; a current. A course of water flowing along a bed in the earth. |
| Experiment | Experiment refers to procedure a scientist uses to study some phenomenon under known conditions. Some experiments are conducted in the laboratory, but others are conducted in nature. The resulting scientific data or facts must be verified or confirmed by repeated observation. |
| Toxic | Harmful, deadly, or poisonous is referred to as toxic. |
| Debris | Debris refers to any accumulation of rock fragments; detritus. |
| Heat | Total kinetic energy of all the randomly moving atoms, ions, or molecules within a given substance, excluding the overall motion of the whole object. This form of kinetic energy flows from one body to another when there is a temperature difference betwe is referred to as heat. |
| Competition | Two or more individual organisms of a single species or two or more individuals of different species attempting to use the same scarce resources in the same ecosystem is a competition. |

# Chapter 13. Antipredator Behavior

## Chapter 13. Antipredator Behavior

| | |
|---|---|
| Diurnal | Diurnal refers to literally of the day, but here meaning having a period or a tidal day, i.e. about 24.84 hours. |
| Brittle | Behavior of material where stress causes abrupt fracture are called the brittle. |
| Probability | A mathematical statement about how likely it is that something will happen is probability. |
| Density | Density refers to the ratio of a mass to a unit volume specified as grams per cubic centimeter. |
| Peninsula | An elongated portion of land nearly surrounded by water and connected to a larger body of land, usually by a neck or an isthmus is called a peninsula. |
| Response | The amount of health damage caused by exposure to a certain dose of a harmful substance or form of radiation is a response. |
| Reproduction | Production of offspring by one is called reproduction. |
| Benefits | Benefits refers to the economic value of a scheme, usually measured in terms of the cost of damages avoided by the scheme, or the valuation of perceived amenity or environmental improvements. |
| Key | Key refers to a low, insular BANK of sand, coral, etc., as one of the islets off the southern coast of Florida. |
| Acid | A substance that releases a hydrogen ion in solution is an acid. |
| Morphology | Morphology refers to river/estuary/lake/seabed form and its change with time. |
| Front | The boundary between two air masses with different temperatures and densitie is referred to as front. |
| Polychaete | A class of annelid worms that includes most marine segmented worms is called polychaete. |
| Intertidal zone | The area of shoreline between low and high tides are called the intertidal zone. |
| Zone | Division or province of the ocean with homogeneous characteristics is referred to as a zone. |
| Disperse | Disperse refers to to spread or distribute from a fixed or constant source. To cause to become widely separated. |
| Energy | Capacity to do work by performing mechanical, physical, chemical, or electrical tasks or to cause a heat transfer between two objects at different temperatures is an energy. |
| Tests | The skeleton or shells of certain microorganisms are called tests. |
| Vertebrates | Vertebrates refer to animals with backbones. Compare with invertebrates. |
| Invertebrates | Invertebrates refers to animals that have no backbones. Compare vertebrates. |
| Sea | Sea refers to the ocean. A large body of salt water, second in rank to an ocean, more or less landlocked and generally part of, or connected with, an ocean or a larger sea. State of the ocean or lake surface, in regard to waves. |
| Duration | In forecasting waves, the length of time the wind blows in essentially the same direction over the FETCH is a duration. |
| Conditions | Conditions refers to physical or chemical attributes of the environment that, while not being consumed, influence biological processes and population growth. Examples are temperature, salinity, and acidity. Compare resources. |
| Fracture | Fracture refers to a general term for any breaks in rock. Fractures include faults, joints, and crack,_. |
| Fat | Fat refers to an organic compound composed of carbon, hydrogen, and oxygen that is insoluble |

# Chapter 13. Antipredator Behavior

## Chapter 13. Antipredator Behavior

| | |
|---|---|
| | in water. |
| Reserves | Reserves refer to resources that have been identified from which a usable mineral can be extracted profitably at present prices with current mining. |
| Bluff | A high, steep BANK or CLIFF is referred to as bluff. |
| Tongue | A long narrow strip of land, projecting into a body of water is a tongue. |
| Discharge | The volume of water flowing in a stream per unit of time is the discharge. |
| Accuracy | Accuracy refers to the extent to which a measurement agrees with the accepted or correct value for that quantity, based on careful measurements by many people over a long time. Compare to precision. |
| Disturbance | Disturbance refers to a discrete event in time that disrupts an ecosystem or community. Examples of natural disturbances include fires, hurricanes, tornadoes, droughts, and floods. Examples of humancaused disturbances include deforestation, overgrazing, and plowing. |
| Temperature | Temperature refers to a measure of the average speed of motion of the atoms, ions, or molecules in a substance or combination of substances at a given moment. Compare with heat. |
| Mixture | Mixture refers to combination of two or more elements and compounds. |
| Cohesion | Attachment of water molecules to each other by hydrogen bonds is a cohesion. |
| Host | Plant or animal on which a parasite feeds is referred to as host. |
| Monitoring | Monitoring refers to the process of collecting data on a regular basis at specific sites to provide a database from which to evaluate change. |
| Growth rate | The net increase in some factor per unit time. In ecology, the growth rate of a population is sometimes measured as the increase in numbers of individuals or biomass per unit time and sometimes as a percentage increase in numbers or biomass per unit time. |
| Reduce | With respect to waste management, reduce refers to practices that will reduce the amount of waste we produce. |
| Hypothesis | In science, an explanation set forth in a manner that can be tested and is capable of being disproved. A tested hypothesis is accepted until and unless it has been disproved. |
| Schooling | The tendency of many species of fishes and mammals to organize themselves into groups is referred to as schooling. |
| Efficiency | The ratio of output to input. With machines, usually the ratio of work or power produced to the energy or power used to operate or fuel them. With living things, efficiency may be defined as either the useful work done or the energy stored in a useful fo. |

# Chapter 13. Antipredator Behavior

## Chapter 14. Sexual Selection

| | |
|---|---|
| Interference | Interference refers to during succession, one species prevents the entrance of later successional species into an ecosystem. For example, some grasses produce such dense and thick mats that seeds of trees cannot reach the soil to germinate. As long as these grasses persist, th. |
| Force | Force refers to a push or pull that affects motion. The product of mass and acceleration of a material. |
| Species | Group of organisms that resemble one another in appearance, behavior, chemical makeup and processes, and genetic structure is a species. Organisms that reproduce sexually are classified as members of the same species only if they can breed with one another and produce offspring. |
| Competition | Two or more individual organisms of a single species or two or more individuals of different species attempting to use the same scarce resources in the same ecosystem is a competition. |
| Well | Well refers to a hole, generally cylindrical and usually walled or lined with pipe, that is dug or drilled into the ground to penetrate an aquifer below the zone of saturation. |
| Natural selection | Process by which a particular beneficial gene is reproduced more than other genes in succeeding generations is natural selection. The result of natural selection is a population that contains a greater proportion of organisms better adapted to certain environments. |
| Spoils | Unwanted rock and other waste materials produced when a material is removed from the earth's surface or subsurface by mining, dredging, quarrying, or excavation are called spoils. |
| Lead | Lead refers to a heavy metal that is an important constituent of automobile batteries and other industrial products. A toxic metal capable of causing environmental disruption and producing a health problem to people and other living organisms. |
| Kingdom | Kingdom refers to the largest category of biological classification. Five kingdoms are presently recognized. |
| Head | A comparatively high promontory with either a CLIFF or steep face. It extends into a large body of water, such as a sea or lake. An unnamed HEAD is usually called a headland. The section of RIP CURRENT which has widened out seaward of the BREAKERS, also called head of |
| Front | The boundary between two air masses with different temperatures and densitie is referred to as front. |
| Resource | Resource refers to anything obtained from the living and nonliving environment to meet human needs and wants. It can also be applied to other species. |
| Theory | A general explanation of a characteristic of nature consistently supported by observation or experiment is referred to as a theory. |
| Hypothesis | In science, an explanation set forth in a manner that can be tested and is capable of being disproved. A tested hypothesis is accepted until and unless it has been disproved. |
| Degree | An arbitrary measure of temperature. One degree Celsius _ 1.8 degrees Fahrenheit. |
| Reproduction | Production of offspring by one is called reproduction. |
| Food | General term for organic molecules capable of providing energy to heterotrophs when combined with oxygen during biochemical respiration is called food. |
| Accumulation | Buildup of matter, energy, or information in a system is referred to as accumulation. |
| Point | Point refers to the extreme end of a cape, or the outer end of any land area protruding into the water, usually less prominent than a cape. A low profile shoreline promontory of more or less triangular shape, the top of which extends seaward. |

## Chapter 14. Sexual Selection

## Chapter 14. Sexual Selection

| | |
|---|---|
| Predation | Predation refers to a situation in which an organism of one species captures and feeds on parts or all of an organism of another species. |
| Observations | Information obtained through one or more of the five senses or through instruments that extend the senses are observations. |
| Hierarchy | Hierarchy refers to grouping of objects by degrees of complexity, grade, or class. A hierarchical system of nomenclature is based on distinctions within groups and between groups. |
| Threshold | Threshold refers to a point in the operation of a system at which a change occurs. With respect to toxicology, it is a level below which effects are not observable and above which effects become apparent. |
| Chain reaction | Multiple nuclear fissions, taking place within a certain mass of a fissionable isotope, that release an enormous amount of energy in a short time is a chain reaction. |
| Premises | In science, initial definitions and assumptions are premises. |
| State | State refers to an expression of the internal form of matter. Water exists in three states: solid, liquid, and gas. A solid has a fixed volume and fixed shape; a liquid has a fixed volume but no fixed shape; and a gas has neither fixed volume nor fixed shape. |
| Spring | A place where groundwater flows out onto the surface is a spring. |
| Population | Group of individual organisms of the same species living within a particular area is referred to as a population. |
| Host | Plant or animal on which a parasite feeds is referred to as host. |
| Probability | A mathematical statement about how likely it is that something will happen is probability. |
| Coast | A strip of land of indefinite length and width that extends from the SEASHORE inland to the first major change in terrain features is referred to as coast. |
| Ventral | Ventral refers to of or pertaining to the underside of animals. |
| Forest | Forest refers to biome with enough average annual precipitation to support growth of various species of trees and smaller forms of vegetation. Compare desert, grassland. |
| Site | A factor considering the summation of all environmental features of a location that influences the placement of a city is a site. |
| Molt | The process of dispensing with an exoskeleton in order to secrete a bigger exoskeleton that will accommodate a larger body size is referred to as molt. |
| System | A set of components that function and interact in some regular and theoretically predictable manner is called a system. |
| Reduce | With respect to waste management, reduce refers to practices that will reduce the amount of waste we produce. |
| Fertility | Fertility refers to the ability to produce offspring; the proportion of births to population. |
| Forcing | With respect to global change, processes capable of changing global temperature, such as changes in solar energy emitted from the sun, or volcanic activity, we have forcing. |
| Taxa | Taxa refer to categories that identify groups of living organisms based upon evolutionary relationships or similarity of characters. |
| Viscous | Ease of flow is viscous. The more viscous a substance, the less readily it flows. |
| Genetic diversity | Genetic diversity refers to variability in the genetic makeup among individuals within a single species. |

Go to **Cram101.com** for the Practice Tests for this Chapter.

# Chapter 14. Sexual Selection

## Chapter 14. Sexual Selection

| | |
|---|---|
| Benefits | Benefits refers to the economic value of a scheme, usually measured in terms of the cost of damages avoided by the scheme, or the valuation of perceived amenity or environmental improvements. |
| Range | Land used for grazing is referred to as the range. |
| Energy | Capacity to do work by performing mechanical, physical, chemical, or electrical tasks or to cause a heat transfer between two objects at different temperatures is an energy. |
| Conditions | Conditions refers to physical or chemical attributes of the environment that, while not being consumed, influence biological processes and population growth. Examples are temperature, salinity, and acidity. Compare resources. |
| Dna | Large molecules in the cells of organisms that carry genetic information in living organisms are dna. |
| Genes | Genes refers to coded units of information about specific traits that are passed on from parents to offspring during reproduction. They consist of segments of DNA molecules found in chromosomes. |
| Resources | Resources refer to substances that can be consumed by an organism and, as a result, become unavailable to other organisms. |
| Base | A substance that combines with a hydrogen ion in solution is called the base. |
| Output | Output refers to matter, energy, or information leaving a system. Compare with input, throughput. |
| Prey | Organism that is captured and serves as a source of food for an organism of another species is called prey. |
| Nutrients | Nutrients refer to chemicals such as phosphorus and nitrogen that, when released into water sources, may cause pollution events such as eutrophication. |
| Fall | Fall refers to a mass moving nearly vertical and downward under the influence of gravity. |
| Plants | Eukaryotic, mostly multicelled organisms such as algae, mosses, ferns, flowers, cacti, grasses, beans, wheat, rice, and trees are plants. These organisms use photosynthesis to produce organic nutrients for themselves and for other organisms is referred to as plants. |
| Micronutrients | Micronutrients refers to chemical elements organisms need in small or even trace amounts to live, grow, or reproduce. Compare with macronutrients. |
| Ion | Ion refers to atom or group of atoms with one or more positive or negative electrical charges. Compare atone, molecule. |
| Mud | Mud refers to a mixture of silt and clay sized particles. |
| Deep water | Deep water in regard to waves, where DEPTH is greater than one-half the WAVE LENGTH. Deep-water conditions are said to exist when the surf waves are not affected by conditions on the bottom. |
| Desert | Desert refers to biome in which evaporation exceeds precipitation and the average amount of precipitation is less than 25 centimeters a year. Such areas have little vegetation or have widely spaced, mostly low vegetation. Compare forest, grassland. |
| Detritus | Small fragments of rock which have been worn or broken away from a mass by the action of water or waves are referred to as detritus. |
| Relative humidity | Relative humidity refers to the amount of water vapor in a certain mass of air, expressed as a percentage of the maximum amount it could hold at that temperature. Compare with absolute humidity. |

# Chapter 14. Sexual Selection

## Chapter 14. Sexual Selection

| | |
|---|---|
| Humidity | A measure of the amount of water vapor in an air mass is a humidity. |
| Sand | Sand refers to an unconsolidated mixture of inorganic soil consisting of small but easily distinguishable grains ranging in size from about .062 mm to 2.0 mm. |
| Development | Development refers to change from a society that is largely rural, agricultural, illiterate, and poor, with a rapidly growing population, to one that is mostly urban, industrial, educated, and wealthy, with a slowly growing or stationary population. |
| Reach | Reach refers to an arm of the ocean extending into the land. A straight section of restricted waterway of considerable extent; may be similar to a narrows, except much longer in extent. |
| Tests | The skeleton or shells of certain microorganisms are called tests. |
| Mass | The amount of material in an object is the mass. |
| Depth | Depth refers to vertical distance from still-water level to the bottom. |
| Environment | Environment refers to all external conditions and factors, living and nonliving, that affect an organism or other specified system during its lifetime; the earth's life-support systems for us and for all other forms of life-another term for solar capita. |
| Parasite | Consumer organism that lives on or in and feeds on a living plant or animal, known as the host, over an extended period of time is a parasite. The parasite draws nourishment from and gradually weakens its host; it may or may not kill the host. |
| Load | The quantity of sediment transported by a current. It includes the suspended load of small particles in the water, and the bedload of large particles that move along the bottom. |
| Risk | Risk refers to the probability that something undesirable will happen from deliberate or accidental exposure. |
| Response | The amount of health damage caused by exposure to a certain dose of a harmful substance or form of radiation is a response. |
| Dorsal | The upper or back surface of an animal is a dorsal. |
| Gene | Gene refers to the fundamental unit in inheritance; it carries the characteristics of parents to their offspring. |
| Alleles | Slightly different molecular forms found in a particular gene are alleles. |
| Caudal fin | The tail of a fish is referred to as caudal fin. |
| Strain | A change in torn or size of a body due to external forces is called strain. |
| Duration | In forecasting waves, the length of time the wind blows in essentially the same direction over the FETCH is a duration. |
| Current | Current refers to the flowing of water, or other liquid or gas. That portion of a stream of water which is moving with a velocity much greater than the average or in which the progress of the water is principally concentrated. Ocean currents can be classified in a number of different ways. |
| Recent | A synonym of Holocene is called recent. |

## Chapter 14. Sexual Selection

## Chapter 15. Parental Care and Mating Systems

| | |
|---|---|
| Resources | Resources refer to substances that can be consumed by an organism and, as a result, become unavailable to other organisms. |
| Species | Group of organisms that resemble one another in appearance, behavior, chemical makeup and processes, and genetic structure is a species. Organisms that reproduce sexually are classified as members of the same species only if they can breed with one another and produce offspring. |
| Genes | Genes refers to coded units of information about specific traits that are passed on from parents to offspring during reproduction. They consist of segments of DNA molecules found in chromosomes. |
| Reproduction | Production of offspring by one is called reproduction. |
| Event | Event refers to an occurrence meeting specified conditions, e.g. damage, a threshold wave height or a threshold water level. |
| Lead | Lead refers to a heavy metal that is an important constituent of automobile batteries and other industrial products. A toxic metal capable of causing environmental disruption and producing a health problem to people and other living organisms. |
| Current | Current refers to the flowing of water, or other liquid or gas. That portion of a stream of water which is moving with a velocity much greater than the average or in which the progress of the water is principally concentrated. Ocean currents can be classified in a number of different ways. |
| Hypothesis | In science, an explanation set forth in a manner that can be tested and is capable of being disproved. A tested hypothesis is accepted until and unless it has been disproved. |
| Food | General term for organic molecules capable of providing energy to heterotrophs when combined with oxygen during biochemical respiration is called food. |
| Oil | The liquid form of petroleum consisting of a complex mixture of large hydrocarbon molecules is referred to as oil. |
| State | State refers to an expression of the internal form of matter. Water exists in three states: solid, liquid, and gas. A solid has a fixed volume and fixed shape; a liquid has a fixed volume but no fixed shape; and a gas has neither fixed volume nor fixed shape. |
| Bore | A steep wave that moves upriver during the flooding tide is the bore. |
| Well | Well refers to a hole, generally cylindrical and usually walled or lined with pipe, that is dug or drilled into the ground to penetrate an aquifer below the zone of saturation. |
| Energy | Capacity to do work by performing mechanical, physical, chemical, or electrical tasks or to cause a heat transfer between two objects at different temperatures is an energy. |
| Reach | Reach refers to an arm of the ocean extending into the land. A straight section of restricted waterway of considerable extent; may be similar to a narrows, except much longer in extent. |
| Adaptation | Any genetically controlled structural, physiological, or behavioral characteristic that helps an organism survive and reproduce under a given set of environmental conditions. It usually results from a beneficial mutation is an adaptation. |
| Competition | Two or more individual organisms of a single species or two or more individuals of different species attempting to use the same scarce resources in the same ecosystem is a competition. |
| Nourishment | Nourishment refers to the process of replenishing a beach. It may be brought about naturally, by longshore transport, or artificially by the deposition of dredged materials. |
| Conditions | Conditions refers to physical or chemical attributes of the environment that, while not being consumed, influence biological processes and population growth. Examples are temperature, |

Go to **Cram101.com** for the Practice Tests for this Chapter.

## Chapter 15. Parental Care and Mating Systems

## Chapter 15. Parental Care and Mating Systems

| | |
|---|---|
| | salinity, and acidity. Compare resources. |
| Response | The amount of health damage caused by exposure to a certain dose of a harmful substance or form of radiation is a response. |
| Fall | Fall refers to a mass moving nearly vertical and downward under the influence of gravity. |
| Risk | Risk refers to the probability that something undesirable will happen from deliberate or accidental exposure. |
| Population | Group of individual organisms of the same species living within a particular area is referred to as a population. |
| Predation | Predation refers to a situation in which an organism of one species captures and feeds on parts or all of an organism of another species. |
| Desert | Desert refers to biome in which evaporation exceeds precipitation and the average amount of precipitation is less than 25 centimeters a year. Such areas have little vegetation or have widely spaced, mostly low vegetation. Compare forest, grassland. |
| Natural selection | Process by which a particular beneficial gene is reproduced more than other genes in succeeding generations is natural selection. The result of natural selection is a population that contains a greater proportion of organisms better adapted to certain environments. |
| Forcing | With respect to global change, processes capable of changing global temperature, such as changes in solar energy emitted from the sun, or volcanic activity, we have forcing. |
| Point | Point refers to the extreme end of a cape, or the outer end of any land area protruding into the water, usually less prominent than a cape. A low profile shoreline promontory of more or less triangular shape, the top of which extends seaward. |
| Vertebrates | Vertebrates refer to animals with backbones. Compare with invertebrates. |
| Frequency | Number of events in a given time interval. For earthquakes, it is the number of cycles of seismic waves that pass in a second; frequency = l/period. |
| Lake | Large natural body of standing fresh water formed when water from precipitation, land runoff, or groundwater flow fills a depression in the earth created by glaciation, earth movement, volcanic activity, or a giant meteorit are called the lake. |
| Predator | Predator refers to an organism that captures and feeds on parts or all of an organism of another species. |
| Aquatic | Aquatic refers to pertaining to water. |
| Development | Development refers to change from a society that is largely rural, agricultural, illiterate, and poor, with a rapidly growing population, to one that is mostly urban, industrial, educated, and wealthy, with a slowly growing or stationary population. |
| Environment | Environment refers to all external conditions and factors, living and nonliving, that affect an organism or other specified system during its lifetime; the earth's life-support systems for us and for all other forms of life-another term for solar capita. |
| Resource | Resource refers to anything obtained from the living and nonliving environment to meet human needs and wants. It can also be applied to other species. |
| Interface | The surface or boundary at which two different substances are in contact, for example, the air-sea interface. |
| Observations | Information obtained through one or more of the five senses or through instruments that extend the senses are observations. |
| Population | Major abiotic and biotic factors that tend to increase or decrease the population size and |

# Chapter 15. Parental Care and Mating Systems

## Chapter 15. Parental Care and Mating Systems

| | |
|---|---|
| dynamics | the age and sex composition of a species is referred to as population dynamics. |
| System | A set of components that function and interact in some regular and theoretically predictable manner is called a system. |
| Site | A factor considering the summation of all environmental features of a location that influences the placement of a city is a site. |
| Parasitism | Interaction between species in which one organism, called the parasite, preys on another organism, called the host, by living on or in the host is referred to as parasitism. |
| Situation | The relative geographic location of a site that makes it a good location for a city is a situation. |
| Benefits | Benefits refers to the economic value of a scheme, usually measured in terms of the cost of damages avoided by the scheme, or the valuation of perceived amenity or environmental improvements. |
| Host | Plant or animal on which a parasite feeds is referred to as host. |
| Head | A comparatively high promontory with either a CLIFF or steep face. It extends into a large body of water, such as a sea or lake. An unnamed HEAD is usually called a headland. The section of RIP CURRENT which has widened out seaward of the BREAKERS, also called head of |
| Reduce | With respect to waste management, reduce refers to practices that will reduce the amount of waste we produce. |
| Coevolution | Evolution in which two or more species interact and exert selective pressures on each other that can lead each species to undergo various adaptation is called coevolution. |
| Parasite | Consumer organism that lives on or in and feeds on a living plant or animal, known as the host, over an extended period of time is a parasite. The parasite draws nourishment from and gradually weakens its host; it may or may not kill the host. |
| Probability | A mathematical statement about how likely it is that something will happen is probability. |
| Duration | In forecasting waves, the length of time the wind blows in essentially the same direction over the FETCH is a duration. |
| Dna | Large molecules in the cells of organisms that carry genetic information in living organisms are dna. |
| Habitat | The place where an organism lives is called habitat. |
| Mud | Mud refers to a mixture of silt and clay sized particles. |
| Degree | An arbitrary measure of temperature. One degree Celsius _ 1.8 degrees Fahrenheit. |
| Heat | Total kinetic energy of all the randomly moving atoms, ions, or molecules within a given substance, excluding the overall motion of the whole object. This form of kinetic energy flows from one body to another when there is a temperature difference betwe is referred to as heat. |
| Dispersion | Dispersion refers to act of dispersing, or state of being dispersed. The separation of waves by virtue of their differing rates of travel. |
| Output | Output refers to matter, energy, or information leaving a system. Compare with input, throughput. |
| Threshold | Threshold refers to a point in the operation of a system at which a change occurs. With respect to toxicology, it is a level below which effects are not observable and above which effects become apparent. |

## Chapter 15. Parental Care and Mating Systems

| | |
|---|---|
| Tests | The skeleton or shells of certain microorganisms are called tests. |
| Coast | A strip of land of indefinite length and width that extends from the SEASHORE inland to the first major change in terrain features is referred to as coast. |
| Forest | Forest refers to biome with enough average annual precipitation to support growth of various species of trees and smaller forms of vegetation. Compare desert, grassland. |
| Range | Land used for grazing is referred to as the range. |
| Annual | Annual refers to plant that grows, sets seed, and dies in one growing season. Compare to perennial. |
| Ecology | Ecology refers to study of the interactions of living organisms with one another and with their nonliving environment of matter and energy; study of the structure and functions of nature. |
| Disperse | Disperse refers to to spread or distribute from a fixed or constant source. To cause to become widely separated. |
| Sounding | Sounding refers to measured depths of water. On hydrographic charts, the soundings are adjusted to a specific plane of reference. |
| Glass | Glass refers to matter created when magma cools too quickly for atoms to arrange themselves into the ordered atomic structures of minerals. Most glasses are supercooled liquids. |
| Reserves | Reserves refer to resources that have been identified from which a usable mineral can be extracted profitably at present prices with current mining. |
| Force | Force refers to a push or pull that affects motion. The product of mass and acceleration of a material. |
| Fertility | Fertility refers to the ability to produce offspring; the proportion of births to population. |
| Resolution | Resolution in general, refers to a measure of the finest detail distinguishable in an object or phenomenon. In particular, a measure of the finest detail distinguishable in an image. |

# Chapter 15. Parental Care and Mating Systems

## Chapter 16. Sociality, Conflict, and Resolution

| | |
|---|---|
| Prey | Organism that is captured and serves as a source of food for an organism of another species is called prey. |
| Food | General term for organic molecules capable of providing energy to heterotrophs when combined with oxygen during biochemical respiration is called food. |
| Energy | Capacity to do work by performing mechanical, physical, chemical, or electrical tasks or to cause a heat transfer between two objects at different temperatures is an energy. |
| Species | Group of organisms that resemble one another in appearance, behavior, chemical makeup and processes, and genetic structure is a species. Organisms that reproduce sexually are classified as members of the same species only if they can breed with one another and produce offspring. |
| Benefits | Benefits refers to the economic value of a scheme, usually measured in terms of the cost of damages avoided by the scheme, or the valuation of perceived amenity or environmental improvements. |
| Well | Well refers to a hole, generally cylindrical and usually walled or lined with pipe, that is dug or drilled into the ground to penetrate an aquifer below the zone of saturation. |
| Core | Core refers to a cylindrical sample extracted from a beach or seabed to investigate the types and DEPTHS of sediment layers. An inner, often much less permeable portion of a BREAKWATER, or BARRIER beach. |
| Theory | A general explanation of a characteristic of nature consistently supported by observation or experiment is referred to as a theory. |
| Resolution | Resolution in general, refers to a measure of the finest detail distinguishable in an object or phenomenon. In particular, a measure of the finest detail distinguishable in an image. |
| Competition | Two or more individual organisms of a single species or two or more individuals of different species attempting to use the same scarce resources in the same ecosystem is a competition. |
| Barrier reef | A coral reef growing around the periphery of an island, but separated from it by a lagoon is referred to as a barrier reef. |
| Reef | A ridge of rock or other material lying just below the surface of the sea is called a reef. |
| Current | Current refers to the flowing of water, or other liquid or gas. That portion of a stream of water which is moving with a velocity much greater than the average or in which the progress of the water is principally concentrated. Ocean currents can be classified in a number of different ways. |
| Environment | Environment refers to all external conditions and factors, living and nonliving, that affect an organism or other specified system during its lifetime; the earth's life-support systems for us and for all other forms of life-another term for solar capita. |
| Point | Point refers to the extreme end of a cape, or the outer end of any land area protruding into the water, usually less prominent than a cape. A low profile shoreline promontory of more or less triangular shape, the top of which extends seaward. |
| Temperature | Temperature refers to a measure of the average speed of motion of the atoms, ions, or molecules in a substance or combination of substances at a given moment. Compare with heat. |
| Hierarchy | Hierarchy refers to grouping of objects by degrees of complexity, grade, or class. A hierarchical system of nomenclature is based on distinctions within groups and between groups. |
| Plankton | Plankton refers to small plant organisms and animal organisms that float in aquatic ecosystems. Compare with benthos, nekton. |

## Chapter 16. Sociality, Conflict, and Resolution

| | |
|---|---|
| Coral | Coral refers to any of more than 6,000 species of small cnidarians, many of which are capable of generating hard calcareous skeletons. |
| Reach | Reach refers to an arm of the ocean extending into the land. A straight section of restricted waterway of considerable extent; may be similar to a narrows, except much longer in extent. |
| Mass | The amount of material in an object is the mass. |
| Parasitism | Interaction between species in which one organism, called the parasite, preys on another organism, called the host, by living on or in the host is referred to as parasitism. |
| Predator | Predator refers to an organism that captures and feeds on parts or all of an organism of another species. |
| Site | A factor considering the summation of all environmental features of a location that influences the placement of a city is a site. |
| Mixture | Mixture refers to combination of two or more elements and compounds. |
| Lead | Lead refers to a heavy metal that is an important constituent of automobile batteries and other industrial products. A toxic metal capable of causing environmental disruption and producing a health problem to people and other living organisms. |
| Situation | The relative geographic location of a site that makes it a good location for a city is a situation. |
| Chemical | One of the millions of different elements and compounds found naturally or synthesized by human is referred to as chemical. |
| Drag | The resistance to movement of an organism induced by the fluid through which it swims is called drag. |
| Range | Land used for grazing is referred to as the range. |
| Recent | A synonym of Holocene is called recent. |
| Conditions | Conditions refers to physical or chemical attributes of the environment that, while not being consumed, influence biological processes and population growth. Examples are temperature, salinity, and acidity. Compare resources. |
| Resources | Resources refer to substances that can be consumed by an organism and, as a result, become unavailable to other organisms. |
| Resource | Resource refers to anything obtained from the living and nonliving environment to meet human needs and wants. It can also be applied to other species. |
| Risk | Risk refers to the probability that something undesirable will happen from deliberate or accidental exposure. |
| Stability | Ability of a living system to withstand or recover from externally imposed changes or stresses is called stability. |
| Observations | Information obtained through one or more of the five senses or through instruments that extend the senses are observations. |
| Experiment | Experiment refers to procedure a scientist uses to study some phenomenon under known conditions. Some experiments are conducted in the laboratory, but others are conducted in nature. The resulting scientific data or facts must be verified or confirmed by repeated observation. |
| Tests | The skeleton or shells of certain microorganisms are called tests. |
| Rings | Rings refers to large whirl-like eddies created by meander cutoffs of strong geostropic |

## Chapter 16. Sociality, Conflict, and Resolution

| | |
|---|---|
| | currents such as the Gulf Stream. They either have warm-water centers or cold-water centers. |
| Reproduction | Production of offspring by one is called reproduction. |
| Mole | In coastal terminology, a massive solid-filled structure of earth, masonry or large stone is referred to as a mole. |
| Population | Group of individual organisms of the same species living within a particular area is referred to as a population. |
| Reduce | With respect to waste management, reduce refers to practices that will reduce the amount of waste we produce. |
| Savanna | Savanna refers to an area with trees scattered widely among dense grasses. |
| Weather | Weather refers to short-term changes in the temperature, barometric pressure, humidity, precipitation, sunshine, cloud cover, wind direction and speed, and other conditions in the troposphere at a given place and time. Compare with climate. |
| Disperse | Disperse refers to to spread or distribute from a fixed or constant source. To cause to become widely separated. |
| State | State refers to an expression of the internal form of matter. Water exists in three states: solid, liquid, and gas. A solid has a fixed volume and fixed shape; a liquid has a fixed volume but no fixed shape; and a gas has neither fixed volume nor fixed shape. |
| Response | The amount of health damage caused by exposure to a certain dose of a harmful substance or form of radiation is a response. |
| Pile | A long substantial pole of wood, concrete or metal, driven into the earth or sea bed to serve as a support or protection is called pile. |
| Degree | An arbitrary measure of temperature. One degree Celsius _ 1.8 degrees Fahrenheit. |
| Algae | Algae refers to simple marine and freshwater plants, unicellular and multicellular, that lack roots, stems, and leaves. |
| Substrate | A general term in reference to the surface on or within which organisms live is a substrate. |
| Habitat | The place where an organism lives is called habitat. |
| Predation | Predation refers to a situation in which an organism of one species captures and feeds on parts or all of an organism of another species. |
| Migration | A term that refers to the habit of some animals is a migration. |
| Fat | Fat refers to an organic compound composed of carbon, hydrogen, and oxygen that is insoluble in water. |
| Reserves | Reserves refer to resources that have been identified from which a usable mineral can be extracted profitably at present prices with current mining. |
| Fuel | Air,/ substar ce that produces heat by combustion is a fuel. |
| Frequency | Number of events in a given time interval. For earthquakes, it is the number of cycles of seismic waves that pass in a second; frequency = l/period. |
| Duration | In forecasting waves, the length of time the wind blows in essentially the same direction over the FETCH is a duration. |
| Head | A comparatively high promontory with either a CLIFF or steep face. It extends into a large body of water, such as a sea or lake. An unnamed HEAD is usually called a headland. The section of RIP CURRENT which has widened out seaward of the BREAKERS, also called head of |

# Chapter 16. Sociality, Conflict, and Resolution

## Chapter 16. Sociality, Conflict, and Resolution

| | |
|---|---|
| **Grassland** | Grassland refers to biome found in regions where moderate annual average precipitation is enough to support the growth of grass and small plants, but not enough to support large stands of trees. Compare desert, forest. |
| **Natural selection** | Process by which a particular beneficial gene is reproduced more than other genes in succeeding generations is natural selection. The result of natural selection is a population that contains a greater proportion of organisms better adapted to certain environments. |
| **Sea** | Sea refers to the ocean. A large body of salt water, second in rank to an ocean, more or less landlocked and generally part of, or connected with, an ocean or a larger sea. State of the ocean or lake surface, in regard to waves. |
| **Shore** | That strip of ground bordering any body of water which is alternately exposed, or covered by tides and/or waves is a shore. A shore of unconsolidated material is usually called a beach. |
| **Molt** | The process of dispensing with an exoskeleton in order to secrete a bigger exoskeleton that will accommodate a larger body size is referred to as molt. |
| **Law** | A large construct explaining events in nature that have been observed to occur with unvarying uniformity under the same conditions is called law. |

# Chapter 16. Sociality, Conflict, and Resolution

| | |
|---|---|
| Recent | A synonym of Holocene is called recent. |
| Species | Group of organisms that resemble one another in appearance, behavior, chemical makeup and processes, and genetic structure is a species. Organisms that reproduce sexually are classified as members of the same species only if they can breed with one another and produce offspring. |
| Predator | Predator refers to an organism that captures and feeds on parts or all of an organism of another species. |
| Food | General term for organic molecules capable of providing energy to heterotrophs when combined with oxygen during biochemical respiration is called food. |
| Hypothesis | In science, an explanation set forth in a manner that can be tested and is capable of being disproved. A tested hypothesis is accepted until and unless it has been disproved. |
| Energy | Capacity to do work by performing mechanical, physical, chemical, or electrical tasks or to cause a heat transfer between two objects at different temperatures is an energy. |
| Reduce | With respect to waste management, reduce refers to practices that will reduce the amount of waste we produce. |
| Benefits | Benefits refers to the economic value of a scheme, usually measured in terms of the cost of damages avoided by the scheme, or the valuation of perceived amenity or environmental improvements. |
| Natural selection | Process by which a particular beneficial gene is reproduced more than other genes in succeeding generations is natural selection. The result of natural selection is a population that contains a greater proportion of organisms better adapted to certain environments. |
| Gene | Gene refers to the fundamental unit in inheritance; it carries the characteristics of parents to their offspring. |
| Alleles | Slightly different molecular forms found in a particular gene are alleles. |
| Population | Group of individual organisms of the same species living within a particular area is referred to as a population. |
| Predation | Predation refers to a situation in which an organism of one species captures and feeds on parts or all of an organism of another species. |
| Risk | Risk refers to the probability that something undesirable will happen from deliberate or accidental exposure. |
| Probability | A mathematical statement about how likely it is that something will happen is probability. |
| Habitat | The place where an organism lives is called habitat. |
| Disperse | Disperse refers to to spread or distribute from a fixed or constant source. To cause to become widely separated. |
| Event | Event refers to an occurrence meeting specified conditions, e.g. damage, a threshold wave height or a threshold water level. |
| Bank | The rising ground bordering a lake, river or sea is referred to as a bank. |
| Desert | Desert refers to biome in which evaporation exceeds precipitation and the average amount of precipitation is less than 25 centimeters a year. Such areas have little vegetation or have widely spaced, mostly low vegetation. Compare forest, grassland. |
| Nourishment | Nourishment refers to the process of replenishing a beach. It may be brought about naturally, by longshore transport, or artificially by the deposition of dredged materials. |

| | |
|---|---|
| Nip | The cut made by waves in a shoreline of emergence is called a nip. |
| Development | Development refers to change from a society that is largely rural, agricultural, illiterate, and poor, with a rapidly growing population, to one that is mostly urban, industrial, educated, and wealthy, with a slowly growing or stationary population. |
| Environment | Environment refers to all external conditions and factors, living and nonliving, that affect an organism or other specified system during its lifetime; the earth's life-support systems for us and for all other forms of life-another term for solar capita. |
| Degree | An arbitrary measure of temperature. One degree Celsius _ 1.8 degrees Fahrenheit. |
| Response | The amount of health damage caused by exposure to a certain dose of a harmful substance or form of radiation is a response. |
| Well | Well refers to a hole, generally cylindrical and usually walled or lined with pipe, that is dug or drilled into the ground to penetrate an aquifer below the zone of saturation. |
| Genes | Genes refers to coded units of information about specific traits that are passed on from parents to offspring during reproduction. They consist of segments of DNA molecules found in chromosomes. |
| Dna | Large molecules in the cells of organisms that carry genetic information in living organisms are dna. |
| Sea | Sea refers to the ocean. A large body of salt water, second in rank to an ocean, more or less landlocked and generally part of, or connected with, an ocean or a larger sea. State of the ocean or lake surface, in regard to waves. |
| Tests | The skeleton or shells of certain microorganisms are called tests. |
| Competition | Two or more individual organisms of a single species or two or more individuals of different species attempting to use the same scarce resources in the same ecosystem is a competition. |
| Hydrocarbons | Compounds containing only carbon and hydrogen are a large group of organic compounds, including petroleum products, such as crude oil and natural gas are hydrocarbons. |
| Plants | Eukaryotic, mostly multicelled organisms such as algae, mosses, ferns, flowers, cacti, grasses, beans, wheat, rice, and trees are plants. These organisms use photosynthesis to produce organic nutrients for themselves and for other organisms is referred to as plants. |
| Mass | The amount of material in an object is the mass. |
| Site | A factor considering the summation of all environmental features of a location that influences the placement of a city is a site. |
| Conditions | Conditions refers to physical or chemical attributes of the environment that, while not being consumed, influence biological processes and population growth. Examples are temperature, salinity, and acidity. Compare resources. |
| Threshold | Threshold refers to a point in the operation of a system at which a change occurs. With respect to toxicology, it is a level below which effects are not observable and above which effects become apparent. |
| System | A set of components that function and interact in some regular and theoretically predictable manner is called a system. |
| Fire | The rapid combination of oxygen with organic material to produce flame, heat, and light is referred to as the fire. |
| Pest | Unwanted organism that directly or indirectly interferes with human activities is referred to as a pest. |

| | |
|---|---|
| Chemical | One of the millions of different elements and compounds found naturally or synthesized by human is referred to as chemical. |
| Theory | A general explanation of a characteristic of nature consistently supported by observation or experiment is referred to as a theory. |
| Lead | Lead refers to a heavy metal that is an important constituent of automobile batteries and other industrial products. A toxic metal capable of causing environmental disruption and producing a health problem to people and other living organisms. |
| Experiment | Experiment refers to procedure a scientist uses to study some phenomenon under known conditions. Some experiments are conducted in the laboratory, but others are conducted in nature. The resulting scientific data or facts must be verified or confirmed by repeated observation. |
| Point | Point refers to the extreme end of a cape, or the outer end of any land area protruding into the water, usually less prominent than a cape. A low profile shoreline promontory of more or less triangular shape, the top of which extends seaward. |
| Prey | Organism that is captured and serves as a source of food for an organism of another species is called prey. |
| Frequency | Number of events in a given time interval. For earthquakes, it is the number of cycles of seismic waves that pass in a second; frequency = l/period. |
| Force | Force refers to a push or pull that affects motion. The product of mass and acceleration of a material. |
| Terrestrial | Terrestrial refers to pertaining to land. Compare with aquatic. |
| Observations | Information obtained through one or more of the five senses or through instruments that extend the senses are observations. |
| Rock | Any material that makes up a large, natural, continuous part of earth's crust is rock. |
| Stability | Ability of a living system to withstand or recover from externally imposed changes or stresses is called stability. |
| Spring | A place where groundwater flows out onto the surface is a spring. |
| Key | Key refers to a low, insular BANK of sand, coral, etc., as one of the islets off the southern coast of Florida. |
| Hierarchy | Hierarchy refers to grouping of objects by degrees of complexity, grade, or class. A hierarchical system of nomenclature is based on distinctions within groups and between groups. |
| Reproduction | Production of offspring by one is called reproduction. |
| Current | Current refers to the flowing of water, or other liquid or gas. That portion of a stream of water which is moving with a velocity much greater than the average or in which the progress of the water is principally concentrated. Ocean currents can be classified in a number of different ways. |
| Front | The boundary between two air masses with different temperatures and densitie is referred to as front. |
| Lake | Large natural body of standing fresh water formed when water from precipitation, land runoff, or groundwater flow fills a depression in the earth created by glaciation, earth movement, volcanic activity, or a giant meteorit are called the lake. |
| Resource | Resource refers to anything obtained from the living and nonliving environment to meet human |

| | |
|---|---|
| | needs and wants. It can also be applied to other species. |
| Saturation | A chemical state whereby the maximum amount of solute is dissolved under the given conditions is saturation. |
| Density | Density refers to the ratio of a mass to a unit volume specified as grams per cubic centimeter. |
| Gradient | A measure of slope in meters of rise or fall per meter of horizontal distance. More general, a change of a value per unit of distance, e.g. the GRADIENT in longshore transport causes EROSION or ACCRETION. With reference to winds or currents, the rate of increase or decrease in speed, usually in the vertical; or the curve that represents this rate. |
| Coast | A strip of land of indefinite length and width that extends from the SEASHORE inland to the first major change in terrain features is referred to as coast. |
| Annual | Annual refers to plant that grows, sets seed, and dies in one growing season. Compare to perennial. |
| Rift valley | Rift valley refers to the fault-bounded valley found along the crest of many ocean ridges; it is created by tensional stresses that accompany the process of sea-floor spreading. |
| Valley | An elongated depression, usually with an outlet, between bluffs or between ranges of hills or mountains is a valley. |
| Adaptation | Any genetically controlled structural, physiological, or behavioral characteristic that helps an organism survive and reproduce under a given set of environmental conditions. It usually results from a beneficial mutation is an adaptation. |
| Invertebrates | Invertebrates refers to animals that have no backbones. Compare vertebrates. |
| Soil | Soil refers to a layer of weathered, unconsolidated material on top of bedrock; often also defined as containing organic matter and being capable of supporting plant growth. |
| Organism | Any form of life is referred to as an organism. |
| Entrance | The entrance to a navigable BAY, HARBOR or CHANNEL, INLET or mouth separating the ocean from an inland water body. |
| Debris | Debris refers to any accumulation of rock fragments; detritus. |
| Chromosome | A grouping of various genes and associated proteins in plant and animal cells that carry certain types of genetic informatio are called the chromosome. |
| Cell | Smallest living unit of an organism. Each cell is encased in an outer membrane or wall and contains genetic material and other parts to perform its life function. Organisms such as bacteria consist of only one cell, but most of the organisms we are. |
| Input | Input refers to matter, energy, or information entering a system. Compare output, throughput. |
| Drag | The resistance to movement of an organism induced by the fluid through which it swims is called drag. |
| Mutation | A random change in DNA molecules making up genes that can yield changes in anatomy, physiology, or behavior in offspring is a mutation. |
| Larva | The immature form of an animal that differs significantly from the adult form is a larva. |
| Vertebrates | Vertebrates refer to animals with backbones. Compare with invertebrates. |
| Gut | Gut refers to a narrow passage such as a strait or INLET. A CHANNEL in otherwise shallow water, generally formed by water in motion. |
| Symbiotic | Relationships that exist between different organisms that are mutually beneficial are |

| | |
|---|---|
| | referred to as symbiotic. |
| Molt | The process of dispensing with an exoskeleton in order to secrete a bigger exoskeleton that will accommodate a larger body size is referred to as molt. |
| Protein | Protein refers to a family of complex organic compounds containing nitrogen and composed of various amino acids. |
| Nutrients | Nutrients refer to chemicals such as phosphorus and nitrogen that, when released into water sources, may cause pollution events such as eutrophication. |
| Situation | The relative geographic location of a site that makes it a good location for a city is a situation. |
| Parasitism | Interaction between species in which one organism, called the parasite, preys on another organism, called the host, by living on or in the host is referred to as parasitism. |
| Head | A comparatively high promontory with either a CLIFF or steep face. It extends into a large body of water, such as a sea or lake. An unnamed HEAD is usually called a headland. The section of RIP CURRENT which has widened out seaward of the BREAKERS, also called head of |
| Gene pool | Gene pool refers to the sum total of all genes found in the individuals of the population of a particular species. |
| Pool | Common bed form produced by scour in meandering and straight channels is a pool. |

## Chapter 18. Maintaining Group Cohesion

| | |
|---|---|
| Cohesion | Attachment of water molecules to each other by hydrogen bonds is a cohesion. |
| Chemical | One of the millions of different elements and compounds found naturally or synthesized by human is referred to as chemical. |
| Substrate | A general term in reference to the surface on or within which organisms live is a substrate. |
| Species | Group of organisms that resemble one another in appearance, behavior, chemical makeup and processes, and genetic structure is a species. Organisms that reproduce sexually are classified as members of the same species only if they can breed with one another and produce offspring. |
| Tundra | The treeless land area in alpine and arctic areas characterized by plants of low stature and including bare areas without any plants and covered areas with lichens, mosses, grasses, sedges, and small flowering plants, including low shrubs are referred to as the tundra. |
| Basin | A large submarine depression of a generally circular, elliptical or oval shape is called a basin. |
| Genes | Genes refers to coded units of information about specific traits that are passed on from parents to offspring during reproduction. They consist of segments of DNA molecules found in chromosomes. |
| Resources | Resources refer to substances that can be consumed by an organism and, as a result, become unavailable to other organisms. |
| Nourishment | Nourishment refers to the process of replenishing a beach. It may be brought about naturally, by longshore transport, or artificially by the deposition of dredged materials. |
| Situation | The relative geographic location of a site that makes it a good location for a city is a situation. |
| Conditions | Conditions refers to physical or chemical attributes of the environment that, while not being consumed, influence biological processes and population growth. Examples are temperature, salinity, and acidity. Compare resources. |
| Response | The amount of health damage caused by exposure to a certain dose of a harmful substance or form of radiation is a response. |
| Population | Group of individual organisms of the same species living within a particular area is referred to as a population. |
| Recent | A synonym of Holocene is called recent. |
| Benefits | Benefits refers to the economic value of a scheme, usually measured in terms of the cost of damages avoided by the scheme, or the valuation of perceived amenity or environmental improvements. |
| Force | Force refers to a push or pull that affects motion. The product of mass and acceleration of a material. |
| Habitat | The place where an organism lives is called habitat. |
| Well | Well refers to a hole, generally cylindrical and usually walled or lined with pipe, that is dug or drilled into the ground to penetrate an aquifer below the zone of saturation. |
| Fog | Vapor condensed to fine particles of water and obscuring vision near the ground is a fog. |
| Depth | Depth refers to vertical distance from still-water level to the bottom. |
| Sea | Sea refers to the ocean. A large body of salt water, second in rank to an ocean, more or less landlocked and generally part of, or connected with, an ocean or a larger sea. State of the ocean or lake surface, in regard to waves. |

## Chapter 18. Maintaining Group Cohesion

## Chapter 18. Maintaining Group Cohesion

| | |
|---|---|
| Head | A comparatively high promontory with either a CLIFF or steep face. It extends into a large body of water, such as a sea or lake. An unnamed HEAD is usually called a headland. The section of RIP CURRENT which has widened out seaward of the BREAKERS, also called head of |
| Range | Land used for grazing is referred to as the range. |
| Acceleration | Acceleration refers to cause to move faster. The rate of change of motion. |
| Fold | Fold refers to wavy geologic structures formed by the compression and bending of sedimentary layers. |
| State | State refers to an expression of the internal form of matter. Water exists in three states: solid, liquid, and gas. A solid has a fixed volume and fixed shape; a liquid has a fixed volume but no fixed shape; and a gas has neither fixed volume nor fixed shape. |
| Frequency | Number of events in a given time interval. For earthquakes, it is the number of cycles of seismic waves that pass in a second; frequency = l/period. |
| Amplitude | Half of the peak-to-trough range of a wave is the amplitude. |
| Environment | Environment refers to all external conditions and factors, living and nonliving, that affect an organism or other specified system during its lifetime; the earth's life-support systems for us and for all other forms of life-another term for solar capita. |
| Exoskeleton | Exoskeleton refers to a skeleton partially or completely covering the exterior of a plant or animal. |
| Atoms | Atoms refers to minute units made of subatomic particles that are the basic building blocks of all chemical elements and thus all matter; the smallest unit of an element that can exist and still have the unique characteristics of that element. Compare to ion, molecule. |
| Molecule | Molecule refers to a combination of two or more atoms of the same chemical element or different chemical elements held together by chemical bonds. Compare with atom, ion. |
| Volatile | Substances that readily become gases when pressure is decreased, or temperature increased are referred to as volatile. |
| Concentration | Amount of a chemical in a particular volume or weight of air, water, soil, or other medium is referred to as concentration. |
| Gradient | A measure of slope in meters of rise or fall per meter of horizontal distance. More general, a change of a value per unit of distance, e.g. the GRADIENT in longshore transport causes EROSION or ACCRETION. With reference to winds or currents, the rate of increase or decrease in speed, usually in the vertical; or the curve that represents this rate. |
| Mixture | Mixture refers to combination of two or more elements and compounds. |
| Acid | A substance that releases a hydrogen ion in solution is an acid. |
| Food | General term for organic molecules capable of providing energy to heterotrophs when combined with oxygen during biochemical respiration is called food. |
| System | A set of components that function and interact in some regular and theoretically predictable manner is called a system. |
| Reproduction | Production of offspring by one is called reproduction. |
| Reach | Reach refers to an arm of the ocean extending into the land. A straight section of restricted waterway of considerable extent; may be similar to a narrows, except much longer in extent. |
| Tongue | A long narrow strip of land, projecting into a body of water is a tongue. |
| Wind | Wind refers to the mass movement of air. |

## Chapter 18. Maintaining Group Cohesion

## Chapter 18. Maintaining Group Cohesion

| | |
|---|---|
| Vertebrates | Vertebrates refer to animals with backbones. Compare with invertebrates. |
| Competition | Two or more individual organisms of a single species or two or more individuals of different species attempting to use the same scarce resources in the same ecosystem is a competition. |
| Maria | Dark, low-lying areas of the Moon is called Maria. |
| Site | A factor considering the summation of all environmental features of a location that influences the placement of a city is a site. |
| Duration | In forecasting waves, the length of time the wind blows in essentially the same direction over the FETCH is a duration. |
| Seismic | Referring to vibrations in the Earth produced by earthquakes is referred to as seismic. |
| Ripple | Ripple refers to the light fretting or ruffling on the surface of the water caused by a breeze. The smallest class of waves and one in which the force of restoration is, to a significant degree, both surface tension and gravity. |
| Wave | An oscillatory movement in a body of water manifested by an alternate rise and fall of the surface is called a wave. Disturbances of the surface of a liquid body, as the ocean, in the form of a ridge, swell or hump. The term wave by itself usually refers to the term surface gravity wave. |
| Cell | Smallest living unit of an organism. Each cell is encased in an outer membrane or wall and contains genetic material and other parts to perform its life function. Organisms such as bacteria consist of only one cell, but most of the organisms we are. |
| Current | Current refers to the flowing of water, or other liquid or gas. That portion of a stream of water which is moving with a velocity much greater than the average or in which the progress of the water is principally concentrated. Ocean currents can be classified in a number of different ways. |
| Discharge | The volume of water flowing in a stream per unit of time is the discharge. |
| Matter | Matter refers to anything that has mass and takes up space. On earth, where gravity is present, we weigh an object to determine its mass. |
| Reduce | With respect to waste management, reduce refers to practices that will reduce the amount of waste we produce. |
| Schooling | The tendency of many species of fishes and mammals to organize themselves into groups is referred to as schooling. |
| Calm | The condition of the water surface when there is no WIND WAVES or SWELL is called calm. |
| Sexual reproduction | Sexual reproduction refers to reproduction in organisms that produce offspring by combining sex cells or gametes from both parents. This produces offspring that have combinations of traits from their parents. Compare with asexual reproduction. |
| Energy | Capacity to do work by performing mechanical, physical, chemical, or electrical tasks or to cause a heat transfer between two objects at different temperatures is an energy. |
| Density | Density refers to the ratio of a mass to a unit volume specified as grams per cubic centimeter. |
| Power | Power refers to the time rate of doing work. |
| Zone | Division or province of the ocean with homogeneous characteristics is referred to as a zone. |
| Experiment | Experiment refers to procedure a scientist uses to study some phenomenon under known conditions. Some experiments are conducted in the laboratory, but others are conducted in nature. The resulting scientific data or facts must be verified or confirmed by repeated |

## Chapter 18. Maintaining Group Cohesion

## Chapter 18. Maintaining Group Cohesion

| | |
|---|---|
| | observation. |
| Probability | A mathematical statement about how likely it is that something will happen is probability. |
| Predator | Predator refers to an organism that captures and feeds on parts or all of an organism of another species. |
| Divergence | Divergence refers to to move apart from a common source. |
| Convergence | Resemblance among species belonging to different taxonomic groups as the result from adaptation to similar environments is the convergence. |
| Disperse | Disperse refers to to spread or distribute from a fixed or constant source. To cause to become widely separated. |
| Natural selection | Process by which a particular beneficial gene is reproduced more than other genes in succeeding generations is natural selection. The result of natural selection is a population that contains a greater proportion of organisms better adapted to certain environments. |
| Resource | Resource refers to anything obtained from the living and nonliving environment to meet human needs and wants. It can also be applied to other species. |
| Migration | A term that refers to the habit of some animals is a migration. |
| Lead | Lead refers to a heavy metal that is an important constituent of automobile batteries and other industrial products. A toxic metal capable of causing environmental disruption and producing a health problem to people and other living organisms. |
| Observations | Information obtained through one or more of the five senses or through instruments that extend the senses are observations. |
| Loop | That part of a STANDING WAVE where the vertical motion is greatest and the horizontal velocities are least is a loop. |
| Gravity | The attraction between bodies of matter is a gravity. |
| Horizon | Horizon refers to the line or circle which forms the apparent boundary between Earth and sky. A plane in rock strata characterized by particular features, as occurrence of distinctive fossil species. One of the series of distinctive layers found in a vertical cross-section of any well-developed soil. |
| Point | Point refers to the extreme end of a cape, or the outer end of any land area protruding into the water, usually less prominent than a cape. A low profile shoreline promontory of more or less triangular shape, the top of which extends seaward. |
| Hypothesis | In science, an explanation set forth in a manner that can be tested and is capable of being disproved. A tested hypothesis is accepted until and unless it has been disproved. |
| Plates | Various-sized areas of earth's lithosphere that move slowly around with the mantle's flowing asthenosphere are plates. Most earthquakes and volcanoes occur around the boundaries of these plates. |
| Plate | One of about a dozen rigid segments of Earth's lithosphere that move independently is a plate. The plate consists of continental or oceanic crust and the cool, rigid upper mantle directly below the crust. |
| Accuracy | Accuracy refers to the extent to which a measurement agrees with the accepted or correct value for that quantity, based on careful measurements by many people over a long time. Compare to precision. |
| Oil | The liquid form of petroleum consisting of a complex mixture of large hydrocarbon molecules is referred to as oil. |

Go to **Cram101.com** for the Practice Tests for this Chapter.

# Chapter 18. Maintaining Group Cohesion

## Chapter 18. Maintaining Group Cohesion

| | |
|---|---|
| **Blade** | Blade refers to algal equivalent of a vascular plant's leaf. Also called a frond. |
| **Monitoring** | Monitoring refers to the process of collecting data on a regular basis at specific sites to provide a database from which to evaluate change. |
| **Tests** | The skeleton or shells of certain microorganisms are called tests. |
| **Spring** | A place where groundwater flows out onto the surface is a spring. |
| **Event** | Event refers to an occurrence meeting specified conditions, e.g. damage, a threshold wave height or a threshold water level. |
| **Organism** | Any form of life is referred to as an organism. |
| **Invertebrates** | Invertebrates refers to animals that have no backbones. Compare vertebrates. |
| **Tension** | A state of stress that tends to pull the body apart is called tension. |

## Chapter 19. Maintaining Group Cohesion: The Evolution of Communication

| | |
|---|---|
| Cohesion | Attachment of water molecules to each other by hydrogen bonds is a cohesion. |
| Species | Group of organisms that resemble one another in appearance, behavior, chemical makeup and processes, and genetic structure is a species. Organisms that reproduce sexually are classified as members of the same species only if they can breed with one another and produce offspring. |
| System | A set of components that function and interact in some regular and theoretically predictable manner is called a system. |
| Benefits | Benefits refers to the economic value of a scheme, usually measured in terms of the cost of damages avoided by the scheme, or the valuation of perceived amenity or environmental improvements. |
| State | State refers to an expression of the internal form of matter. Water exists in three states: solid, liquid, and gas. A solid has a fixed volume and fixed shape; a liquid has a fixed volume but no fixed shape; and a gas has neither fixed volume nor fixed shape. |
| Prey | Organism that is captured and serves as a source of food for an organism of another species is called prey. |
| Food | General term for organic molecules capable of providing energy to heterotrophs when combined with oxygen during biochemical respiration is called food. |
| Bluff | A high, steep BANK or CLIFF is referred to as bluff. |
| Resources | Resources refer to substances that can be consumed by an organism and, as a result, become unavailable to other organisms. |
| Population | Group of individual organisms of the same species living within a particular area is referred to as a population. |
| Mutation | A random change in DNA molecules making up genes that can yield changes in anatomy, physiology, or behavior in offspring is a mutation. |
| Preservation | Static protection of an area or element, attempting to perpetuate the existence of a given 'state' is called preservation. |
| Genes | Genes refers to coded units of information about specific traits that are passed on from parents to offspring during reproduction. They consist of segments of DNA molecules found in chromosomes. |
| Force | Force refers to a push or pull that affects motion. The product of mass and acceleration of a material. |
| Response | The amount of health damage caused by exposure to a certain dose of a harmful substance or form of radiation is a response. |
| Situation | The relative geographic location of a site that makes it a good location for a city is a situation. |
| Hypothesis | In science, an explanation set forth in a manner that can be tested and is capable of being disproved. A tested hypothesis is accepted until and unless it has been disproved. |
| Energy | Capacity to do work by performing mechanical, physical, chemical, or electrical tasks or to cause a heat transfer between two objects at different temperatures is an energy. |
| Risk | Risk refers to the probability that something undesirable will happen from deliberate or accidental exposure. |
| Well | Well refers to a hole, generally cylindrical and usually walled or lined with pipe, that is dug or drilled into the ground to penetrate an aquifer below the zone of saturation. |

# Chapter 19. Maintaining Group Cohesion: The Evolution of Communication

## Chapter 19. Maintaining Group Cohesion: The Evolution of Communication

| | |
|---|---|
| Environment | Environment refers to all external conditions and factors, living and nonliving, that affect an organism or other specified system during its lifetime; the earth's life-support systems for us and for all other forms of life-another term for solar capita. |
| Degree | An arbitrary measure of temperature. One degree Celsius _ 1.8 degrees Fahrenheit. |
| Predation | Predation refers to a situation in which an organism of one species captures and feeds on parts or all of an organism of another species. |
| Reach | Reach refers to an arm of the ocean extending into the land. A straight section of restricted waterway of considerable extent; may be similar to a narrows, except much longer in extent. |
| Resource | Resource refers to anything obtained from the living and nonliving environment to meet human needs and wants. It can also be applied to other species. |
| Head | A comparatively high promontory with either a CLIFF or steep face. It extends into a large body of water, such as a sea or lake. An unnamed HEAD is usually called a headland. The section of RIP CURRENT which has widened out seaward of the BREAKERS, also called head of |
| Frequency | Number of events in a given time interval. For earthquakes, it is the number of cycles of seismic waves that pass in a second; frequency = l/period. |
| Duration | In forecasting waves, the length of time the wind blows in essentially the same direction over the FETCH is a duration. |
| Depression | Depression refers to a general term signifying any depressed or lower area in the ocean floor. |
| Ecology | Ecology refers to study of the interactions of living organisms with one another and with their nonliving environment of matter and energy; study of the structure and functions of nature. |
| Conditions | Conditions refers to physical or chemical attributes of the environment that, while not being consumed, influence biological processes and population growth. Examples are temperature, salinity, and acidity. Compare resources. |
| Marble | Marble refers to metamorphosed limestone. |
| Natural selection | Process by which a particular beneficial gene is reproduced more than other genes in succeeding generations is natural selection. The result of natural selection is a population that contains a greater proportion of organisms better adapted to certain environments. |
| Development | Development refers to change from a society that is largely rural, agricultural, illiterate, and poor, with a rapidly growing population, to one that is mostly urban, industrial, educated, and wealthy, with a slowly growing or stationary population. |
| Point | Point refers to the extreme end of a cape, or the outer end of any land area protruding into the water, usually less prominent than a cape. A low profile shoreline promontory of more or less triangular shape, the top of which extends seaward. |
| Stress | Stress refers to force per unit area. May be compression, tension, or shear. |
| Neck | Neck refers to the narrow strip of land which connects a peninsula with the mainland, or connects two ridges. The narrow band of water flowing seaward through the surf. |
| Swell | Waves that have traveled a long distance from their generating area and have been sorted out by travel into long waves of the same approximate period are referred to as a swell. |
| Temperature | Temperature refers to a measure of the average speed of motion of the atoms, ions, or molecules in a substance or combination of substances at a given moment. Compare with heat. |
| Heat | Total kinetic energy of all the randomly moving atoms, ions, or molecules within a given |

# Chapter 19. Maintaining Group Cohesion: The Evolution of Communication

## Chapter 19. Maintaining Group Cohesion: The Evolution of Communication

| | |
|---|---|
| | substance, excluding the overall motion of the whole object. This form of kinetic energy flows from one body to another when there is a temperature difference betwe is referred to as heat. |
| Plume | Plume refers to an arm of magna rising upward from the mantle. |
| Front | The boundary between two air masses with different temperatures and densitie is referred to as front. |
| Aquatic | Aquatic refers to pertaining to water. |
| Copepods | Members that belong to an order of Crustacea that are shrimplike in appearance and that feed voraciously on phytoplankton are copepods. |
| Habitat | The place where an organism lives is called habitat. |
| Kingdom | Kingdom refers to the largest category of biological classification. Five kingdoms are presently recognized. |
| Predator | Predator refers to an organism that captures and feeds on parts or all of an organism of another species. |
| Chemical | One of the millions of different elements and compounds found naturally or synthesized by human is referred to as chemical. |
| Poison | A chemical that in one dose kills exactly 50% of the animals in a test population within a 14-day period is referred to as poison. |
| Fire | The rapid combination of oxygen with organic material to produce flame, heat, and light is referred to as the fire. |
| Adaptation | Any genetically controlled structural, physiological, or behavioral characteristic that helps an organism survive and reproduce under a given set of environmental conditions. It usually results from a beneficial mutation is an adaptation. |
| Efficiency | The ratio of output to input. With machines, usually the ratio of work or power produced to the energy or power used to operate or fuel them. With living things, efficiency may be defined as either the useful work done or the energy stored in a useful fo. |
| Sea | Sea refers to the ocean. A large body of salt water, second in rank to an ocean, more or less landlocked and generally part of, or connected with, an ocean or a larger sea. State of the ocean or lake surface, in regard to waves. |
| Competition | Two or more individual organisms of a single species or two or more individuals of different species attempting to use the same scarce resources in the same ecosystem is a competition. |
| Ocean basin | Ocean basin refers to deep-ocean floor made of basaltic crust. Compare with continental margin. |
| Basin | A large submarine depression of a generally circular, elliptical or oval shape is called a basin. |
| Climate | Physical properties of the troposphere of an area based on analysis of its weather records over a long period. The two main factors determining an area's climate are temperature, with its seasonal variations, and the amount and distri. |
| Selective pressure | A factor in a population's environment that causes natural selection to occur is selective pressure. |
| Divergence | Divergence refers to to move apart from a common source. |
| Key | Key refers to a low, insular BANK of sand, coral, etc., as one of the islets off the southern coast of Florida. |

# Chapter 19. Maintaining Group Cohesion: The Evolution of Communication

## Chapter 19. Maintaining Group Cohesion: The Evolution of Communication

| | |
|---|---|
| Demand for food | The amount of food that would he bought at a given price if it were available is referred to as demand for food. |
| Reduce | With respect to waste management, reduce refers to practices that will reduce the amount of waste we produce. |
| Precision | Precision refers to a measure of reproducibility, or how closely a series of measurements of the same quantity agree with one another. Compare with accuracy. |
| Lead | Lead refers to a heavy metal that is an important constituent of automobile batteries and other industrial products. A toxic metal capable of causing environmental disruption and producing a health problem to people and other living organisms. |
| Forest | Forest refers to biome with enough average annual precipitation to support growth of various species of trees and smaller forms of vegetation. Compare desert, grassland. |
| Matter | Matter refers to anything that has mass and takes up space. On earth, where gravity is present, we weigh an object to determine its mass. |
| Range | Land used for grazing is referred to as the range. |
| Attenuation | The loss or dissipation of wave energy, resulting in a reduction of wave height is attenuation. |
| Degradation | The geologic process by means of which various parts of the surface of the earth are worn away and their general level lowered, by the action of wind and water is referred to as degradation. |
| Grassland | Grassland refers to biome found in regions where moderate annual average precipitation is enough to support the growth of grass and small plants, but not enough to support large stands of trees. Compare desert, forest. |
| Wind | Wind refers to the mass movement of air. |
| Observations | Information obtained through one or more of the five senses or through instruments that extend the senses are observations. |
| Lake | Large natural body of standing fresh water formed when water from precipitation, land runoff, or groundwater flow fills a depression in the earth created by glaciation, earth movement, volcanic activity, or a giant meteorit are called the lake. |
| Terrace | A horizontal or nearly horizontal natural or artificial topographic feature interrupting a steeper slop, sometimes occurring in a series is called a terrace. |
| Event | Event refers to an occurrence meeting specified conditions, e.g. damage, a threshold wave height or a threshold water level. |
| Tests | The skeleton or shells of certain microorganisms are called tests. |
| Site | A factor considering the summation of all environmental features of a location that influences the placement of a city is a site. |
| Star | A massive sphere of incandescent gases powered by the conversion of hydrogen to helium and other heavier elements is a star. |
| Current | Current refers to the flowing of water, or other liquid or gas. That portion of a stream of water which is moving with a velocity much greater than the average or in which the progress of the water is principally concentrated. Ocean currents can be classified in a number of different ways. |
| Recent | A synonym of Holocene is called recent. |

# Chapter 19. Maintaining Group Cohesion: The Evolution of Communication

Lightning Source UK Ltd.
Milton Keynes UK
UKHW031328240119
336120UK00002B/29/P